Fabrication Techniques and Machining Methods of Advanced Composite Materials

Fabrication Techniques and Machining Methods of Advanced Composite Materials documents the most current inventive developments in the manufacture and machining of sophisticated composite materials. The utilization of cutting-edge engineering materials with exceptional qualities, such as lightweight and long service life, is necessary for the industry to reduce both energy consumption and production/maintenance costs.

It provides scientific and technological insights on the fabrication routes of composites. It covers various applications suitable for the aerospace, nuclear, and medical fields and emphasizes advanced machining techniques. The book also highlights some of the top innovations and advances in the fabrication of advanced composite materials and their processing technologies while targeting the latest applications.

This reference book is meant to be used as a one-stop resource for academics and manufacturing experts, engineers in related fields, and academic researchers. It encapsulates the current trends of today's fabrication and machining processes for advanced composite materials.

Innovations in Smart Manufacturing for Long-Term Development and Growth
Series Editors: Atul Babbar, Gursel Alici, Yu Dong, Ankit Sharma

Fabrication Techniques and Machining Methods of Advanced Composite Materials
Vikas Dhawan, Atul Babbar, Inderdeep Singh, and Jonathan M. Weaver

For more information about this series, please visit: www.routledge.com/Innovations-in-Smart-Manufacturing-for-Long-Term-Development-and-Growth/book-series/CRCISMLDG

Fabrication Techniques and Machining Methods of Advanced Composite Materials

Edited by
Vikas Dhawan
Atul Babbar
Inderdeep Singh
Jonathan M. Weaver

CRC Press
Taylor & Francis Group
Boca Raton London New York

CRC Press is an imprint of the
Taylor & Francis Group, an **informa** business

Designed cover image: Shutterstock

First edition published 2024
by CRC Press
6000 Broken Sound Parkway NW, Suite 300, Boca Raton, FL 33487-2742

and by CRC Press
4 Park Square, Milton Park, Abingdon, Oxon, OX14 4RN

CRC Press is an imprint of Taylor & Francis Group, LLC

© 2024 selection and editorial matter, Vikas Dhawan, Atul Babbar, Inderdeep Singh, and Jonathan M. Weaver; individual chapters, the contributors

ISBN: 978-1-032-53908-9 (hbk)
ISBN: 978-1-032-54843-2 (pbk)
ISBN: 978-1-003-42773-5 (ebk)

DOI: 10.1201/9781003427735

Typeset in Times
by SPi Technologies India Pvt Ltd (Straive)

Contents

Preface

This book aims to present a comprehensive and most recent breakthroughs in the multidisciplinary area of composites. Composites are used extensively in business and daily life in the civilized world as one of the cutting-edge technical materials. However, several mechanical characteristics, such as tensile strength, limit the use of such a material. Various techniques exist now for creating composites. The useful facilities, composite use, composite kind, applied materials, and financial budget rate all play a role in their selection. In essence, each fabrication technique offers benefits and drawbacks. For instance, some processes move quickly but cost a lot, whilst others take a long time yet produce composites at a low cost. For composite materials that are challenging to cut, advanced machining techniques are recommended to minimize these machining impacts. Using power-driven machine tools, machining is the process of removing material from a workpiece. Tools, work materials, and process parameter settings are only a few of the variables that affect how efficiently a process runs and how well the outcome is. The use of composite materials is spreading outside the aerospace industry due to their many advantages in mechanical qualities. The machining of composite materials is difficult. This book becomes a one-stop resource for academics and manufacturing experts, engineers in related fields, academic researchers, etc. by encapsulating the current trends of today's fabrication and machining processes for advanced composite materials. Hence, we are highly confident that this contribution will benefit all the readers in different aspects.

Editors

Dr. Vikas Dhawan is an academician, administrator, and leader in the field of higher education having more than 25 years of rich experience. He has administrative experience in the capacity of pro-vice chancellor, principal, additional director, director and director principal in various institutes of repute. He has filed 35 patents and has more than 50 research publications in international and national journals and conferences of repute. His main areas of research are machining and testing of fibre-reinforced composites and he has guided two Ph.D. students in the field of biodegradable composites. He has organized many conferences, seminars, FDPs and workshops and has chaired sessions in international and national conferences. He has established many labs, Incubation centre and centres of excellence in the field of Mechanical Engineering. He is also trained in the field of Pneumatics by SMC, India, and in the field of Entrepreneurship by DST and NIESBUD. He has keen interest in the use of technology and innovation in teaching and implementation of New Education Policy (2020). He likes to lead from the front and is skilled in team building, asset management, resource allocation and FOOP control.

Dr. Atul Babbar works as deputy dean (research and development) and an assistant professor at SGT University, Haryana, India. He has completed his Ph.D. from Thapar Institute of Engineering and Technology, Patiala, India. During his Ph.D., he worked on neurosurgical bone grinding. He has been teaching and guiding students in academics and the research field. His research interests include interdisciplinary research, particularly where engineering and medical research intersect. Apart from that, his research is focused on medical devices, manufacturing, hybrid machining processes, biomedical and additive manufacturing, and not limited to that. He has authored more than 60 research articles in various International and National Web of Science (SCI/SCIE), and Scopus-indexed journals. His research papers have over 800+ citations with an h-index-17 and i10 index of 31. He has 20 Published Indian patents and 04 Granted International patents. He is serving as a guest editor for various reputed journals indexed in the Web of Science, PubMed, and Scopus databases. He is also serving as a book series editor for Taylor & Francis Group. He has reviewed research articles for various peer-reviewed Web of Science and Scopus-indexed journals. He has published many international books and book chapters in multiple databases. He is an associate member of the Institution of Engineers (AMIE), India, and a professional member of the Association for Computing Machinery.

Dr. Inderdeep Singh working as a professor in the Department of Mechanical and industrial engineering at Indian Institute of Technology (IIT) Roorkee. He has completed his doctoral from Indian Institute of Technology (IIT), Roorkee. He has authored many research articles in various International/National Web of Science (SCI/SCIE), and Scopus indexed journals. He has reviewed articles for many international and national peer-review journals. He has supervised many post graduates and doctoral students. He is a member of Asian-Australasian Association on Composite Materials.

Dr. Jonathan M. Weaver is currently working as a professor in the Department of Mechanical Engineering at University of Detroit Mercy, United States. He has published many articles in the various high quality peer reviewed journals. He has reviewed articles for many international and national peer-review journals. He has supervised many post graduates and doctoral students. He is a certified professional member of ICAgile. He teaches statics, dynamics, vehicle dynamics, robotics, machine design, DOE, mechanical measurements laboratory, CAE, innovation, product planning and development, systems engineering and systems architecture. His research interests relate to robotics, vehicle dynamics, design of experiments, robust design, innovation and creativity, and the product development process.

Contributors

Atul Babbar
Mechanical Engineering Department,
   Shree Guru Gobind Singh
   Tricentenary University
Gurugram, India

Sandeep Bansal
Mechanical Engineering Department,
   Shree Guru Gobind Singh
   Tricentenary University
Gurugram, India

Harpreet Kaur Channi
Department of Electrical Engineering,
   Chandigarh University
Mohali, India

Vikas Dhawan
Mechanical Engineering Department,
   Shree Guru Gobind Singh
   Tricentenary University
Gurugram, India

Pallav Gupta
Amity School of Engineering and
   Technology, Amity University
Noida, India

Yaman Hooda
Manav Rachna International Institute of
   Research and Studies
Faridabad, India

Harish Kumar
Department of Mechanical Engineering,
   National Institute of Technology
   (NITD)
Delhi, India

Rajender Kumar
Mechanical Engineering Department,
   School of Engineering & Technology,
   Manav Rachna International Institute
   of Research & Studies
Faridabad, Haryana, India

Raman Kumar
Department of Mechanical and
   Production Engineering, Guru Nanak
   Dev Engineering College
Ludhiana, India

Vidyapati Kumar
Department of Mechanical Engineering,
   Indian Institute of Technology
Kharagpur, India
Faridabad, India

Nelson Laishram
Mechanical Engineering Department,
   Shree Guru Gobind Singh
   Tricentenary University
Gurugram, India

Jimmy Mehta
Manav Rachna International Institute of
   Research and Studies
Faridabad, India

Ankita Mistri
Department of Mechanical Engineering,
   Indian Institute of Technology
Dhanbad, India

Prateek Mittal
Manav Rachna International Institute of
   Research and Studies
Faridabad, India

K. Ponappa
Mechanical Engineering Department,
   Indian Institute of Information
   Technology, Design, and
   Manufacturing
Jabalpur, Madhya Pradesh, India

Ankit Sharma
Chitkara University Institute of
   Engineering & Technology, Chitkara
   University
Rajpura, India

Vikas Sharma
Department of Mechanical Engineering,
  National Institute of Technical
  Teachers' Training and Research
  (NITTTR)
Chandigarh, India

Lavish Kumar Singh
Department of Mechanical Engineering,
  Sharda University
Greater Noida, India

Manpreet Singh
Department of Mechanical Engineering,
  Chitkara University Institute of
  Engineering and Technology
Rajpura, India

Okram Greatson Singh
Mechanical Engineering Department,
  Shree Guru Gobind Singh
  Tricentenary University
Gurugram, India

Ramandeep Singh Sidhu
Department of Mechanical and
  Production Engineering, Guru Nanak
  Dev Engineering College
Ludhiana, India

P. Sudhakar Rao
Department of Mechanical Engineering,
  National Institute of Technical
  Teachers' Training and Research
  (NITTTR)
Chandigarh, India

Sri Phani Sushma
Department of Mechanical
  Engineering, University College
  of Engineering Narasaraopet,
  JNTU Kakinada, India

Puneet Tandon
Mechanical Engineering Department,
  Indian Institute of Information
  Technology, Design, and
  Manufacturing
Jabalpur, Madhya Pradesh, India

Ankit Tiwari
Mechanical Engineering Department,
  Indian Institute of Information
  Technology, Design, and
  Manufacturing
Jabalpur, Madhya Pradesh, India

1 The State of the Art in Advanced Fabrication and Characterization Techniques for Composites
Research Insights and Future Directions

Vidyapati Kumar
Indian Institute of Technology, Kharagpur, India

Ankita Mistri
Indian Institute of Technology, Dhanbad, India

Atul Babbar
Shree Guru Gobind Singh Tricentenary University,
Gurugram, India

Ankit Sharma
Chitkara University Institute of Engineering & Technology,
Chitkara University, Rajpura, India

*Sandeep Bansal, Nelson Laishram,
and Okram Greatson Singh*
Shree Guru Gobind Singh Tricentenary University,
Gurugram, India

DOI: 10.1201/9781003427735-1

1.1 INTRODUCTION

Composite materials have evolved as a flexible class of materials with various uses owing to their unique mix of characteristics. Composites consist of two or more mixed components to generate new superior-quality materials. The usage of composites has expanded dramatically in recent years because of its excellent strength-to-weight ratio, corrosion resistance, and durability [1]. The history of composite materials may be traced back to prehistoric times when mud and straw were mixed to build bricks. However, the modern use of composite materials began during World War II when they were used to make aeroplane parts. Since then, composite materials have been extensively employed in different domains, including aerospace, automotive, sports, construction, and medical industries [2, 3].

Advanced manufacturing processes have been developed to manufacture high-performance composite materials with better characteristics. These approaches allow for exact control over the material's microstructure, resulting in better mechanical, thermal, electrical, and chemical characteristics. Some examples of advanced manufacturing processes [4] include 3D printing, automated fibre placement, braiding, filament winding, injection moulding, pultrusion, and resin transfer moulding (RTM). In addition to improved production processes, characterization techniques play a significant role in understanding the characteristics and behaviour of composite materials. Various techniques [5] such as Scanning Electron Microscopy (SEM), Atomic Force Microscopy (AFM), X-Ray Photoelectron Spectroscopy (XPS), Contact Angle Measurement, and Fourier-Transform Infrared Spectroscopy (FTIR) are used for the characterization of composite materials. Composites can be categorized into two primary groups: structural composites and composites with distinct physical or chemical properties determined by their intended applications. These composites are further classified based on the type of matrix utilized, which may be metallic or non-metallic [6]. Non-metallic matrices include polymers, ceramics, and semiconductors. Figure 1.1 visually depicts the classification of composites, showcasing various categories such as ceramics, metals, laminates, particulates, fibrous composites, and polymers.

This chapter offers an overview of the sophisticated production and characterization methods utilized for composites. It will also examine the future possibilities and possible uses of these approaches. The chapter is structured as follows: Section 1.1 introduces composite materials and their manufacturing procedures, while Section 1.2 concentrates on advanced fabrication techniques, including their concepts, applications, benefits, and limits. Section 1.3 examines the characterization methods used for composite materials, their visions, and their applications. Finally, Section 1.4 finishes the chapter by outlining the future possibilities and prospective uses of enhanced manufacturing and characterization methods for composites.

1.2 ADVANCED FABRICATION TECHNIQUES FOR COMPOSITES

Advanced fabrication techniques are an array of manufacturing procedures that employ novel approaches to create composite structures with great precision, accuracy, and repeatability. These approaches have permitted the creation of composites

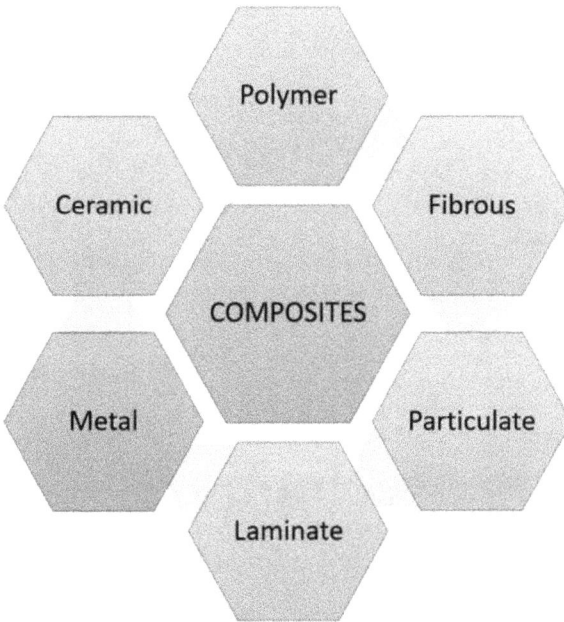

FIGURE 1.1   Categorization of composite materials.

with customized forms, sizes, and characteristics, as well as the incorporation of functional components like sensors, actuators, and electrical conductors [7]. As depicted in Figure 1.2, advanced fabrication methods include additive manufacturing (AM) [8, 9], automated fibre placement (AFP), braiding, filament winding, injection moulding, pultrusion, RTM, and vacuum infusion. Each process has distinct benefits and limits, making it appropriate for various applications and materials.

1.2.1   ADDITIVE MANUFACTURING

Composites have gained significant attention in biomedical applications due to their unique properties, such as high strength-to-weight ratio, biocompatibility, and tailored mechanical properties. AM processes have been utilized to fabricate complex and customized biomedical implants and devices with composite materials.

The AM process involves layer-by-layer material deposition, allowing the creation of complex geometries with precise control over material placement. Various composite materials have been used for AM in biomedical applications, including carbon fibre reinforced polymers, ceramic-polymer composites, and metal-polymer composites [10].

One of the significant challenges in using composites for biomedical applications is ensuring biocompatibility and minimizing the risk of adverse reactions in the human body. Additionally, there is a need for standardized testing and characterization methods to ensure the reliability and safety of the composite materials used in biomedical devices [11, 12].

FIGURE 1.2 Classification of various fabrication techniques for composites.

Despite these challenges, using composites in AM for biomedical applications shows promising potential for creating custom implants and devices to improve patient outcomes. Developing advanced characterization techniques and exploring new composite materials are critical for further advancing the use of composites in biomedical applications.

The typical manufacturing procedure for printing tissues [13] using the AM technique is depicted in Figure 1.3. The first phase in this process is data collection in which the data about the organ or portion of the body that needs to be printed is obtained using medical imaging methods such as X-ray and Computed Tomography (CT) scan. Once the data is received, a 3D model of the component is created using computer-assisted design (CAD) and computer-aided manufacturing (CAM) processes. The bioinks used for 3D bioprinting are then chosen, and the bioprinting settings and resolutions are calibrated. Once the material has been 3D bioprinted, its functionality is tested in an incubator or bioreactor to confirm stability, cell viability, tissue development, and so on. Following these procedures, the printed component is

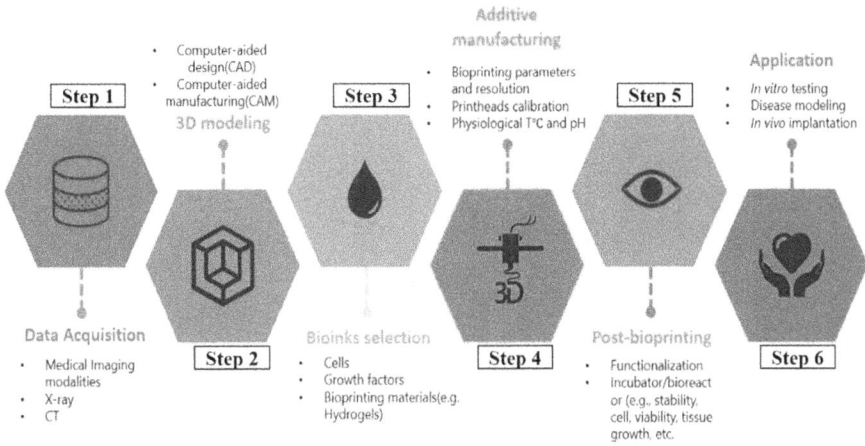

FIGURE 1.3 Typical process layout for additive manufacturing process.

ready for in vitro testing, disease modelling, and in vivo implantation in the human body, saving significant life in healthcare systems.

1.2.2 AUTOMATED FIBRE PLACEMENT

AFP is a cutting-edge manufacturing process extensively utilized in the aerospace and automotive sectors to create complicated composite structures. A computer-controlled placement head is used to accurately lay down a continuous fibre tape on a mandrel or tooling, allowing the fabrication of complex shapes with little material waste.

AFP has a wide range of applications, including the manufacture of aircraft wings, fuselages, and engine nacelles, as well as automobile components and wind turbine blades [14]. The method is ideal for producing big, curved, and complicated structures with high precision and reproducibility. Various high-performance fibres, including carbon, glass, and aramid, and sophisticated resins, such as epoxy and polyester, are utilized in AFP. The materials used are determined by the application's unique needs, such as strength, stiffness, weight, and durability.

AFP has various benefits over alternative manufacturing processes, such as manual lay-up and filament winding, including high efficiency, precision, reproducibility, and automation. Furthermore, the technology reduces material waste and labour costs, resulting in a more environmentally friendly and cost-effective manufacturing process. AFP represents a hybrid manufacturing process that combines the strengths of filament winding and ATL (Automatic Tape Laying) technologies, as depicted in Figure 1.4 [14]. It offers a versatile solution for fabricating both flat and cylindrical structures. Several critical parameters, including feed rate, curing/melting temperature, consolidation force, and lay-up speed, can significantly impact the manufacturing process at three distinct stages (beginning, during, and completion) of the lay-up process. These stages are illustrated in Figure 1.4 [14], and these parameters directly influence the quality of the final laminate.

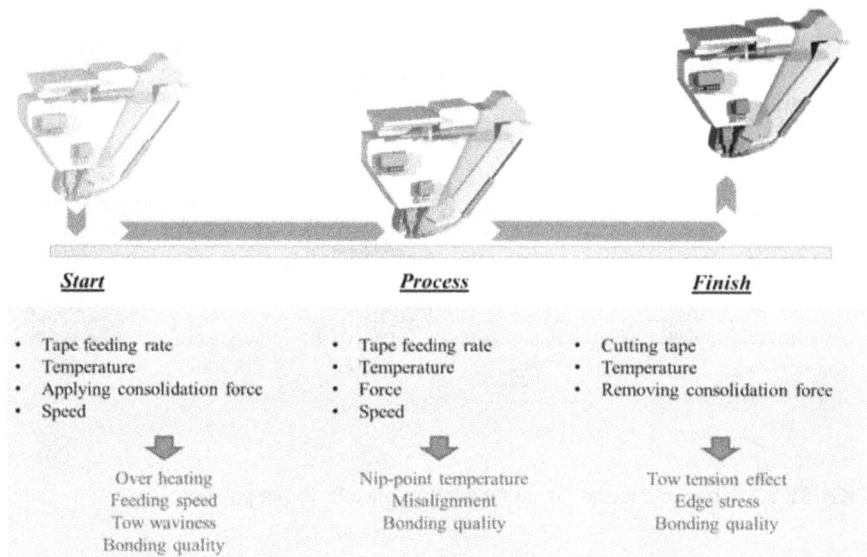

Start	**Process**	**Finish**
• Tape feeding rate	• Tape feeding rate	• Cutting tape
• Temperature	• Temperature	• Temperature
• Applying consolidation force	• Force	• Removing consolidation force
• Speed	• Speed	
Over heating	Nip-point temperature	Tow tension effect
Feeding speed	Misalignment	Edge stress
Tow waviness	Bonding quality	Bonding quality
Bonding quality		

FIGURE 1.4   Process layout and common processing issues during the lay-up in AFP [14].

One of the critical advantages of AFP is its ability to employ multiple tows with narrower widths. This feature enables easier manipulation and placement of the tows, especially on contoured or curved surfaces [15]. Consequently, it becomes more feasible to create multi-stiffened laminates by strategically orienting the fibres in different directions. AFP has been pivotal in transforming composite structure production within the aerospace industry for over two decades. Its implementation has led to notable improvements in quality, accuracy, and overall manufacturing efficiency.

Moreover, AFP has significantly reduced costs by minimizing material waste and optimizing labour utilization [16]. To facilitate a comprehensive comparison, Table 1.1 presents a side-by-side analysis of the main specifications between AFP and conventional composite manufacturing methods [17]. This table offers valuable

TABLE 1.1
Contrasting AFP with Conventional Composite Manufacturing Techniques

Specifications	ATL Approach	AFP Technique	Traditional Methods
Material waste	Minimal	Low waste	Substantial waste
Labour expenses	Affordable	Cost-effective	Expensive
Consistency	High	Reliable	Inconsistent
Precision	High	Accurate	Inaccurate
Efficiency	High	Productive	Low productivity
Cost efficiency	Yes	Yes	No
Variety of materials	Broad range of tapes	Narrow tows	Wide range of tapes
Lay-up speed	Very high	Relatively high	Very high
Components' shape	Large components	Curved & contoured surfaces	Large components

insights into AFP's distinctive features and benefits, aiding in evaluating its suitability for various applications.

The development of novel materials, such as nanocomposites, and the integration of sensors and control systems to allow real-time monitoring and optimization of the production process are among the future objectives for AFP. Furthermore, software and computing power advances are expected to improve AFP's capabilities and expand its applications to other industries such as medicine and sports.

1.2.3 BRAIDING

Braiding is a well-known and commonly used technique for creating complicated composite structures. This method interlaces yarns or fibres at an angle to form a braided design. Braiding has various benefits, including high fibre volume fraction, excellent structural stability, and minimal void content [18]. These characteristics make braided composites suitable for multiple applications, including aerospace, automotive, and sporting goods. There are several braiding procedures, such as axial, bias, and radial braiding. Bias and radial braiding provide high torsional and bending stiffness, whereas axial braiding produces structures with high axial stiffness. The individual application and the required qualities of the finished product determine the braiding procedure used. Figure 1.5 exhibits the braiding process.

Carbon fibres, glass fibres, aramid fibres, and many matrix materials, such as thermoplastics and thermosets, are utilized in braided composites [19]. The final product's intended mechanical, thermal, and chemical qualities determine the materials used. Braiding has various benefits over other composite fabrication processes, including excellent production efficiency, reduced material waste, and the capacity to construct complicated geometries. However, the braiding process has some limitations, such as the difficulty in producing uniform fibre angles, which can lead to variations in mechanical properties. Braiding is projected to be an essential manufacturing technology for advanced composites. Researchers are now looking at methods to increase the efficiency and quality of the braiding process, such as using modern sensors and control systems [20]. Furthermore, braided composites are gaining popularity for applications such as biomedical implants and energy storage devices.

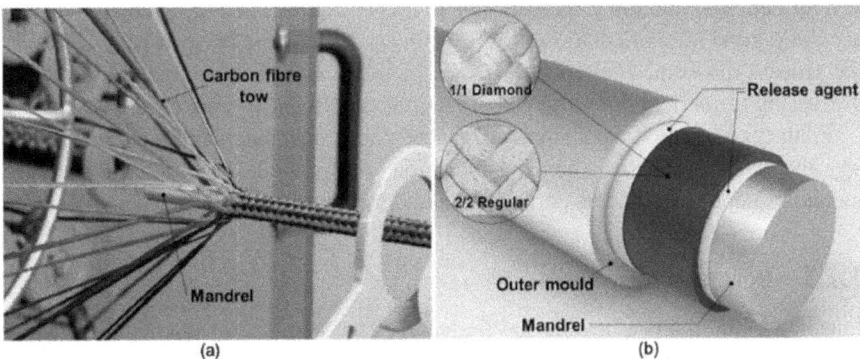

FIGURE 1.5 Schematic representation of the braiding process [19].

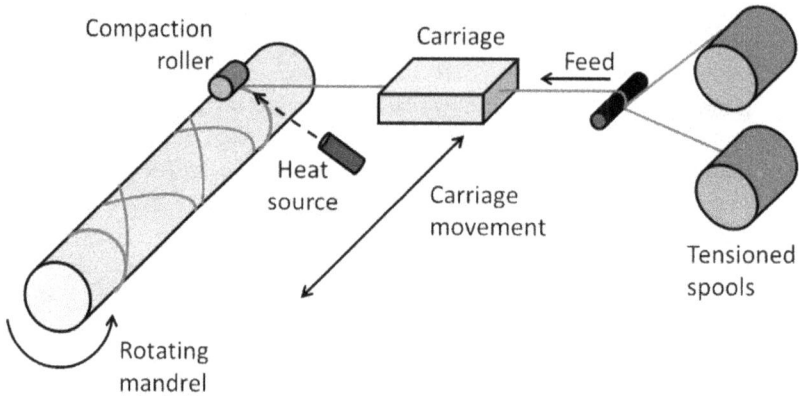

FIGURE 1.6 Working representation of filament winding [21].

1.2.4 FILAMENT WINDING

Another sophisticated composite production technology is filament winding, which continuously wraps resin-impregnated fibres around a rotating mandrel. Due to its capacity to create high-strength, lightweight components, the method is extensively employed in the aerospace, automotive, and sporting goods sectors. A revolving mandrel is covered with a release agent to prevent the composite from adhering [21]. The fibres are then coiled onto the mandrel in a predefined pattern and direction, with tension and winding speed regulated to guarantee proper fibre orientation and coverage. The entire filament winding process is exhibited in Figure 1.6 [21]. The resin is normally administered to the fibres by a resin bath or injection system, ensuring full fibre impregnation.

Filament winding techniques include hoop-and-radial, polar, and cross-wound. The process used is determined by the form and size of the component being manufactured and the mechanical qualities needed [22]. Fibres such as carbon, glass, and aramid are employed in filament winding as are different resin systems such as epoxy, polyester, and vinyl ester. The final product's mechanical qualities and the production circumstances determine the material used. Filament winding provides several benefits, including the capacity to create complicated structures with great accuracy, good fibre orientation and strength, and little waste creation [23]. However, the process has some limitations, such as the need for specialized equipment, long processing times, and difficulties in producing significant components.

Filament winding is a potentially advanced composite material production technology that provides distinct benefits in various applications. Additional research is required to optimize process parameters and develop new materials and applications.

1.2.5 INJECTION MOULDING

Injection moulding is a flexible and frequently used manufacturing method in which molten material is injected into a mould cavity under high pressure and then cooled and hardened to create a completed item [24]. This technology may be customized to produce composite components with complicated forms and internal characteristics, making it a potential solution for various aerospace, automotive, and consumer applications. In composite injection moulding, thermoplastic or thermoset resin is commonly used as the matrix material, reinforced with fibres or particles to improve the mechanical and physical qualities of the final product [25]. The required qualities, cost, and production requirements determine the matrix material and reinforcement to be used.

The composite material is melted and fed into a hot barrel during the injection moulding process. The molten material is pumped into the mould cavity under high pressure through a nozzle [26]. The mould is often constructed of steel or aluminium and is meant to generate the required form of the item. After injection, the mould is cooled to solidify the material, and the component is expelled from the mould cavity. Entire work of this process is exhibited in the visual schematic diagram in Figure 1.7 [26].

One of the benefits of injection moulding is its ability to make components with great dimensional precision and repeatability. The highly automated technique may be modified to high-volume manufacturing with minimal labour costs. Injection moulding can also produce parts with complicated geometries and precise features, making it perfect for applications such as automobile body panels and interior components, electrical enclosures, and medical equipment [27].

However, there are some limitations to the injection moulding process for composites. The high temperatures and pressures required in the process might cause deterioration or damage to the reinforcing fibres, resulting in a loss in the mechanical qualities of the finished product. Furthermore, injection moulding equipment and tooling costs can be relatively high, particularly for small-scale production [27]. Despite these challenges, injection moulding remains a promising technology for composite fabrication, particularly for applications requiring high-volume production, tight tolerances, and complex geometries. Ongoing research is focused on optimizing processing conditions and material choices to optimize the performance of moulded composites.

1.2.6 PULTRUSION

Pultrusion is a continuous manufacturing method that produces high-strength composite materials with a consistent cross-section. The procedure includes pushing continuous fibres through a resin solution and then through a heated die, which cures the resin and consolidates the fibres [28]. The resultant composite is cut to length and finished to the required form. Pultrusion principles and applications are based on continuous fibre reinforcement and thermosetting resins [29]. The method primarily creates structural components with excellent strength-to-weight ratios, stiffness, and corrosion resistance. Typical uses include the building and construction, aircraft, transportation, and maritime sectors. The materials used in pultruded composites might vary based on the application requirements. High-strength fibres such as glass, carbon, or aramid are often used as reinforcement, while thermosetting resins such as epoxy, polyester, or vinyl ester are used as matrix materials.

The procedure starts with a creel that houses spools of unidirectional fibres composed of fibreglass or carbon fibre as shown in Figure 1.8 [30]. These fibres are

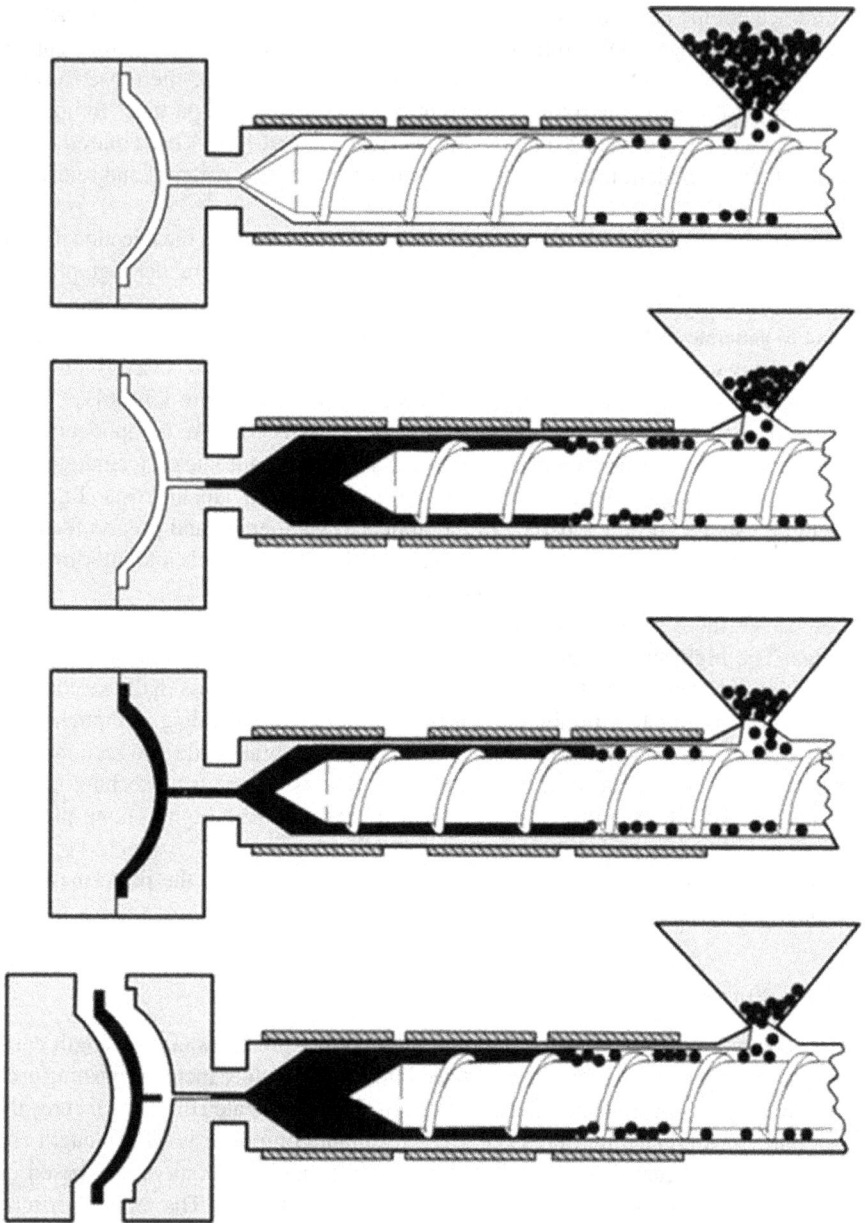

FIGURE 1.7 Schematic representation of the injection moulding process [26].

drawn through a resin bath and impregnated with a liquid thermoset resin, such as polyester or epoxy. After impregnation, the fibres are dragged through a shaping die, which performs the fibres into the appropriate shape and orientation. The prepared fibres are then passed through a succession of heated dies, which cure the thermoset

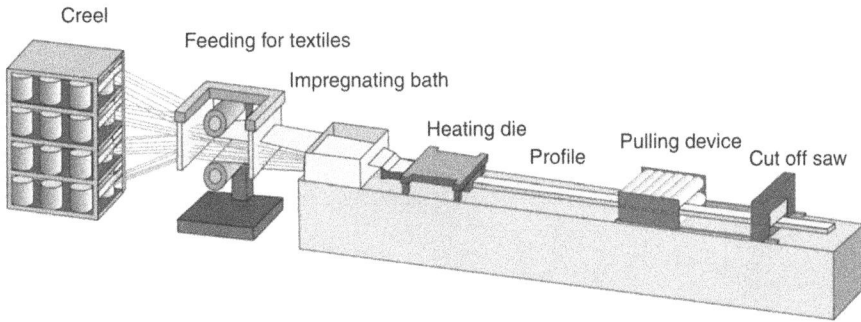

FIGURE 1.8   Schematic overview of a pultrusion device [30].

resin by commencing a chemical process that hardens the material. After curing, the composite material is drawn via a traction system, which maintains consistent tension and guarantees the composite retains its form and dimensions. Finally, the continuous pultruded composite is cut to the desired length and shape with a saw or other cutting tool, as shown in Figure 1.8 [30].

The benefits of pultrusion include a high strength-to-weight ratio, excellent corrosion resistance, and minimal maintenance needs [31]. However, the process is limited to producing constant cross-sections and can be challenging for complex shapes [31]. Furthermore, the high processing temperatures required for curing the resin can restrict the types of fibres and resins used. Overall, pultrusion is an essential advanced manufacturing technology for generating high-strength composite materials with a consistent cross-section. Its application in numerous sectors is expanding, and current research is looking into methods to broaden the spectrum of materials that may be employed while also improving process efficiency.

1.2.7   RESIN TRANSFER MOULDING

RTM is a composite component fabrication technology. RTM injects liquid resin into a closed mould holding dry fibres or preforms, which are infused with resin to produce a solid composite structure [32]. The method is often used to create high-performance items with complicated forms and flawless surface finishes. The pultrusion process involves a sequence of steps, as depicted in Figure 1.9 [33]. The procedure starts with manufacturing preparation, followed by the lay-up and draping of the reinforcement materials. Next, the mould is closed, enabling the injection and curing of resin within the mould cavity. After the resin has solidified, the demoulding and final processing stages take place, preparing the part for removal.

Thermoset resins such as epoxy, polyester, and vinyl ester are often used in RTM, as are different reinforcements such as carbon, glass, and natural fibres. RTM outperforms alternative composite manufacturing techniques in various ways, including the capacity to make oversized, complicated items with outstanding surface finishes and dimensional accuracy [34]. Additionally, the process enables precise control over fibre content and orientation, improving mechanical properties and reducing weight. Furthermore, because RTM is a low-pressure process, it is well suited for producing

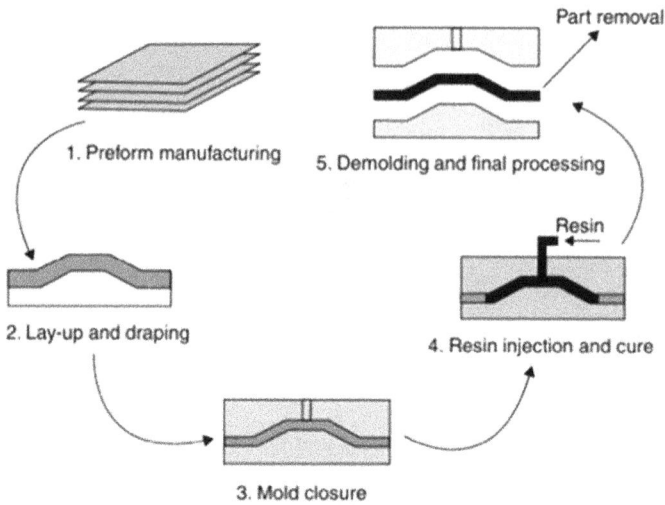

FIGURE 1.9   Procedural steps in resin transfer moulding process [33].

thin-walled or delicate parts. However, RTM has some limitations, including longer cycle times when compared to other methods and the requirement for specialized equipment and tooling. It is also necessary to carefully manage factors like temperature, pressure, and resin flow, which might be difficult [35].

RTM is a popular method for creating high-performance composite products with complicated forms and outstanding surface finishes. Ongoing research focuses on generating novel materials and process enhancements to increase the capabilities and cost-effectiveness of this critical manufacturing technology.

1.2.8   VACUUM INFUSION

Vacuum infusion is a composite manufacturing technique that impregnates dry reinforcing materials with resin using vacuum pressure. This method is often employed to manufacture large, complicated structures requiring high strength-to-weight ratios, including boat hulls, wind turbine blades, and aerospace components [36]. As depicted in Figure 1.10, the procedure starts with placing a dry reinforcing material, such as carbon fibre or fibreglass, into a mould or onto a tool surface. After sealing the mould, a vacuum removes air from the cavity and reinforces the material [37]. Then, gravity or vacuum pressure is used to introduce a resin system into the mould cavity. The resin system is generally a two-part combination of a liquid resin and a hardener that react to generate a solid composite material. The resin impregnates the reinforcing material as it flows through the mould under vacuum pressure, ensuring the structure is appropriately wetted out [37]. The injected composite is then cured using heat and pressure to produce the final product.

Various fibres, including carbon, aramid, and fibreglass, and a number of resins, such as epoxy, polyester, and vinyl ester, are typically used for vacuum infusion. Vacuum infusion has the benefit of producing high-quality composite structures with

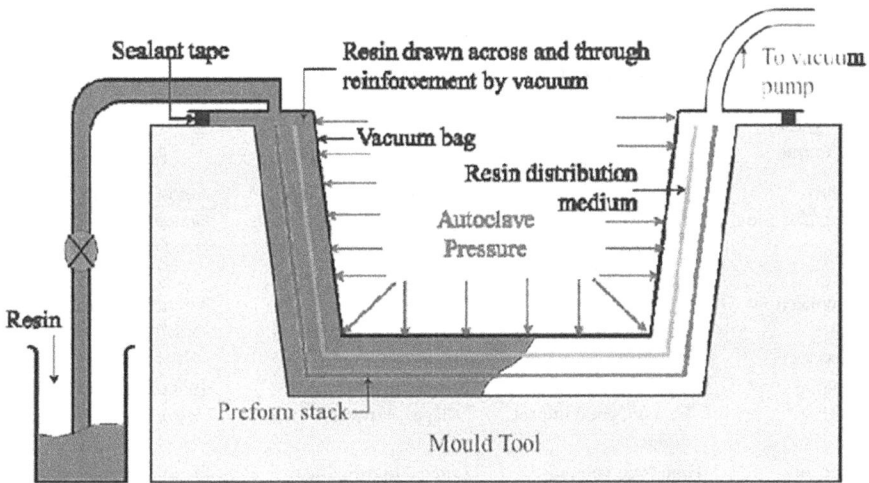

FIGURE 1.10 A schematic illustration of vacuum enhanced resin infusion process [37].

minimal void content and good fibre-to-resin ratios. This method may also be utilized to make complicated shapes and pieces with excellent uniformity and reproducibility. However, the process can be time-consuming and costly, requiring precise control of the resin flow and curing process to avoid defects such as voids or resin-rich areas.

To facilitate a comprehensive understanding, Table 1.2 presents a Comparative Analysis of Fabrication Techniques for Composite Materials. Each technique has been analysed and compared based on their advantages, limitations, and suitability for different applications. This table highlights key factors such as cost, complexity, production speed, material compatibility, and surface finish, allowing researchers and engineers to make informed decisions when selecting the most appropriate fabrication method for their specific requirements.

1.3 CHARACTERIZATION TECHNIQUES FOR COMPOSITES

Surface characterization is a fundamental process that involves analysing a material's surface's physical and chemical properties. It is critical in understanding and optimizing material properties [38, 39] across various applications, including biomedical engineering, electronics, coatings, materials science, and more. In biomedical engineering, surface characterization is indispensable for comprehending and optimizing the properties of implant materials. By meticulously controlling the surface properties of biomedical implants, such as roughness, chemical composition, and topography, it is possible to enhance their biocompatibility, corrosion resistance, wear resistance, and other essential characteristics [40]. However, surface characterization techniques find extensive applications beyond biomedical engineering. For example, in the field of electronics, surface characterization helps assess the quality and performance of electronic components, ensuring proper functionality and reliability. In the coatings industry, surface characterization techniques enable the analysis of coating thickness, adhesion, and surface roughness to ensure optimal coating performance.

TABLE 1.2

Comparative Analysis of Fabrication Techniques for Composite Materials

Fabrication Technique	Advantages	Disadvantages	Applications
Additive Manufacturing	Ability to create complex geometries, customization, reduced waste	Limited materials, low production rate	Aerospace components, automotive parts, medical implants
Automated Fibre Placement	High production rate, precise fibre placement	Limited to flat or moderately curved surfaces	Aircraft wings and fuselages, wind turbine blades, pressure vessels
Braiding	High fibre volume fraction, good impact resistance	Limited to cylindrical shapes, expensive tooling	Tubes and pipes, tennis rackets, golf club shafts
Filament Winding	High fibre volume fraction, precise fibre orientation	Limited to cylindrical shapes, expensive tooling	Pressure vessels, rocket motor cases, golf club shafts
Injection Moulding	High production rate, complex shapes	Limited to short fibre reinforcement, limited material options	Automotive components, electrical housings, consumer products
Pultrusion	Continuous production, high strength-to-weight ratio	Limited to constant cross-sections, expensive tooling	Structural shapes, ladder rails, sporting goods
Resin Transfer Moulding	Good surface finish, low cost tooling	Limited to simple shapes, slow production rate	Marine parts, automotive components, consumer products
Vacuum Infusion	Good surface finish, low cost tooling	Limited to flat or moderately curved shapes, slow production rate	Boat hulls, wind turbine blades, automotive components

Materials science also heavily relies on surface characterization to understand the structure and properties of materials at the atomic and molecular levels [41, 42]. This knowledge aids in developing advanced materials with tailored properties for various industries and applications. Figure 1.11 depicts various surface characterization techniques commonly employed across different fields. These techniques include SEM, AFM, XPS, Contact Angle Measurement, and FTIR. These versatile techniques allow researchers and engineers to gain valuable insights into the surface properties of materials, facilitating advancements in a wide range of applications.

1.3.1 SCANNING ELECTRON MICROSCOPY

SEM is an extensively utilized technique in composite materials. It plays a crucial role in the fabrication and characterization of composites and in assessing their performance in various applications. In composite material fabrication, SEM provides detailed insights into the microstructure and morphology of the constituents, such as fibres, fillers, and matrix materials [43]. It enables researchers and manufacturers to analyse these components' distribution, orientation, and interfacial bonding, which are vital for determining the mechanical properties and overall performance

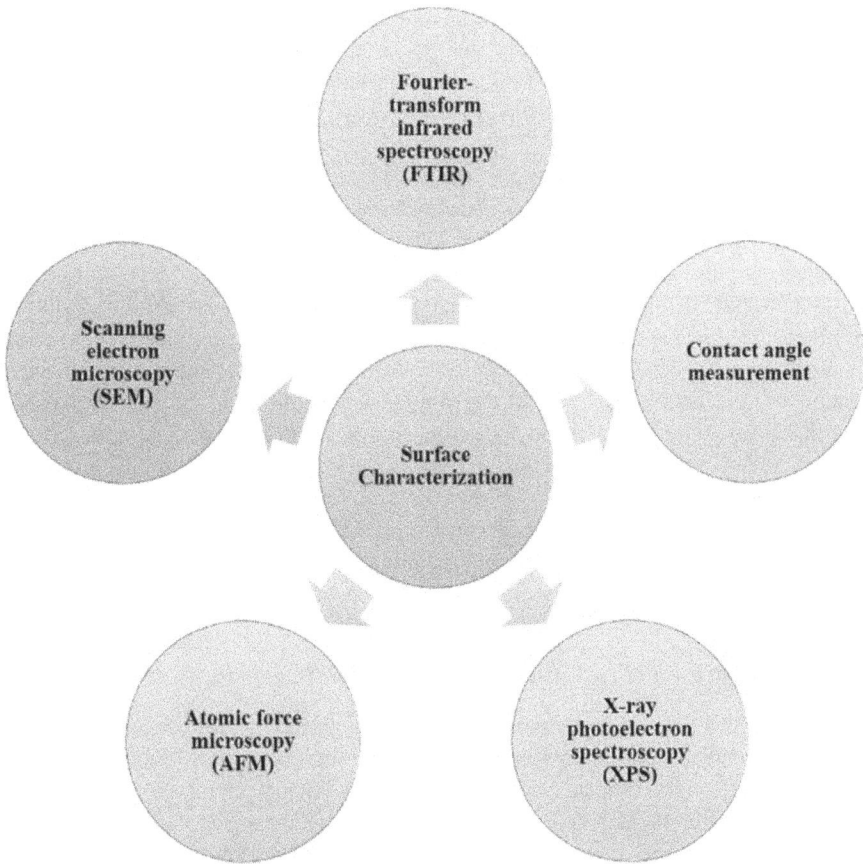

FIGURE 1.11   Various surface characterization techniques.

of composites. Moreover, SEM aids in evaluating the quality and integrity of composite materials by detecting defects, such as voids, cracks, and delamination. This information is invaluable for ensuring the reliability and durability of composites in practical applications [44]. In terms of applications, SEM has wide-ranging significance. In aerospace engineering, SEM helps assess composite aircraft components' structural integrity and damage mechanisms, ensuring their safety and performance under extreme conditions. In automotive manufacturing, SEM is employed to examine the microstructure of composite panels, optimizing their lightweight design and enhancing fuel efficiency. Additionally, SEM contributes to developing advanced composite materials for renewable energy systems, electronic devices, construction materials, and many other industries.

1.3.2 ATOMIC FORCE MICROSCOPY

AFM is a powerful technique that has revolutionized the field of material characterization, including the analysis of composite materials. It offers valuable insights into the surface topography, mechanical properties, and nanoscale features of materials, making it highly relevant in various applications [45]. AFM plays a significant role in the fabrication and assessment of composite materials. It allows researchers and engineers to examine the surface roughness, particle distribution, and fibre arrangement within composites, providing crucial information about their structural integrity and performance [46].

AFM's ability to operate at the nanoscale enables precise measurements of mechanical properties, such as surface stiffness, elasticity, and adhesion forces, which are vital for understanding composite behaviour and optimizing their design. By mapping these properties at a microscopic level, AFM assists in developing composite materials with enhanced mechanical performance and tailored functionalities. Furthermore, AFM contributes to investigating interfacial interactions between composite components, including fibre-matrix interfaces [45]. It facilitates the characterization of interfacial adhesion and the evaluation of composite interfaces' stability and durability, which directly impact composites' overall mechanical strength and reliability. Beyond fabrication and assessment, AFM finds applications in numerous fields. In the electronics industry, AFM enables the imaging and analysis of nanoscale features in electronic devices, aiding in developing miniaturized components. In nanotechnology, AFM allows for manipulating and characterizing nanomaterials, facilitating advancements in nanofabrication and nanoscale research.

1.3.3   X-Ray Photoelectron Spectroscopy

XPS stands as a remarkable technique that holds immense significance in the realm of material analysis, encompassing the characterization of composite materials [47]. By delving into the intricate details of surface chemistry, elemental composition, and chemical states, XPS is a valuable tool with wide-ranging applications [48]. Within the domain of composite materials, XPS assumes a pivotal role in unravelling the mysteries of chemical composition and surface properties. It bestows upon researchers and engineers the power to decipher the elemental constituents present within composites and ascertain their relative concentrations [49]. Moreover, XPS offers valuable insights into these elements' chemical bonding and oxidation states, shedding light on surface functionalities and interfacial interactions that govern composite behaviour. The ability to scrutinize the surface chemistry of composites empowers XPS to optimize material performance and functionality. It opens doors for the analysis of surface modifications, such as coatings or treatments, and their profound impact on the chemical properties of composites. Such knowledge serves as a compass for tailoring composites to specific applications, enhancing their durability, promoting adhesion, and fortifying resistance against the corrosive forces of nature.

Moreover, XPS is prominent in investigating the intricate mechanisms of composite degradation arising from factors like ageing, wear, or exposure to environmental elements. By monitoring chemical changes transpiring on the composite surface, XPS facilitates an understanding of the degradation processes at play. Consequently, this knowledge paves the way for developing strategies to fortify composite longevity and resilience. In addition to its significance in composite materials, XPS finds its niche in a plethora of diverse fields. In semiconductors, it lends its analytical prowess to unravel the mysteries of thin film structures and detect impurities or contaminants. Meanwhile, in the fascinating realm of surface science research, XPS empowers scientists with vital information on surface reactions and the behaviour of catalysts, thus propelling advancements in catalysis and materials synthesis.

1.3.4   Contact Angle Measurement

Contact Angle Measurement, a fascinating technique in surface science, holds immense significance in understanding the wetting behaviour and surface properties of materials, including composite materials. By observing the shape and behaviour of liquid droplets on a solid surface, contact angle measurement unveils valuable insights into interfacial interactions, surface energy, and wettability, making it a versatile tool with diverse applications [1, 50, 51]. Regarding composite materials, contact angle measurement is indispensable for optimizing material performance and tailoring surface characteristics. Researchers and engineers gain crucial information about wetting phenomena, adhesion, and surface tension by measuring the contact angle between a droplet and the composite surface. This knowledge allows fine-tuning composite surfaces to achieve specific wetting properties, such as hydrophobicity or hydrophilicity and facilitates the development of advanced materials for various applications.

Moreover, contact angle measurement is a gateway to understanding interfacial interactions within composites. It offers valuable insights into the adhesion behaviour and compatibility between components, such as fibres and matrices [52]. By studying contact angles at the interface, researchers can assess the wetting properties of the matrix material on the fibre surface, optimizing composite manufacturing processes and improving overall mechanical properties. Beyond composite materials, contact angle measurement finds applications in many fields. In the realm of surface coatings, it assists in designing self-cleaning surfaces and developing coatings with specific wetting characteristics. In biomaterials, contact angle measurement aids in evaluating implant surface biocompatibility and the study of cell-surface interactions [53]. Additionally, in microfluidics, contact angle measurement plays a crucial role in optimizing the performance of fluidic devices, ensuring precise droplet manipulation and efficient flow control.

1.3.5   Fourier-Transform Infrared Spectroscopy

FTIR is an invaluable analytical technique for material characterization, including composite materials. FTIR provides valuable insights into its chemical composition, molecular structure, and functional groups by examining the interaction between infrared radiation and a material, making it a versatile tool with diverse applications [54–56].

FTIR plays a pivotal role in unravelling the chemical makeup and identifying the specific functional groups present in composite materials. By analysing the absorption and transmission of infrared light at different wavelengths, FTIR spectra offer a fingerprint of the composite's molecular vibrations, enabling researchers to identify the types of bonds and chemical species within the material. This knowledge provides critical information about composite fabrication's composition, polymerization, and curing processes [57]. Moreover, FTIR allows for the assessment of chemical changes and interactions that occur within composite materials. It enables the identification of chemical reactions, such as cross-linking or degradation, providing insights into the stability and durability of composites under various environmental conditions. By monitoring the changes in FTIR spectra over time, researchers can evaluate composite materials' performance and ageing characteristics, facilitating the development of improved formulations and manufacturing processes. FTIR spectroscopy also proves valuable in the analysis of composite surfaces and interfaces. By performing attenuated total reflection (ATR) measurements, FTIR can probe the chemical composition, and functional groups present at the surface, shedding light on surface modifications, coatings, or treatments. This information aids in understanding adhesion phenomena, interfacial interactions, and the optimization of surface properties for enhanced performance in diverse applications [57].

Beyond composite materials, FTIR finds extensive applications in numerous fields. In pharmaceuticals, it assists in drug formulation analysis, stability studies, and identification of impurities. In the field of environmental science, FTIR aids in the identification and quantification of pollutants in air, water, and soil samples. Additionally, in forensic analysis, FTIR is a powerful tool for identifying unknown substances, such as drugs or trace evidence.

1.4 CONCLUSION AND FUTURE DIRECTIONS

The field of composites has witnessed significant advancements in fabrication and characterization techniques, paving the way for exciting developments and innovations in the future. Advanced manufacturing processes such as 3D printing, automated fibre placement, braiding, filament winding, injection moulding, pultrusion, and RTM have revolutionized the production of composites, creating complex and high-performance structures with enhanced efficiency and precision. Moreover, characterization techniques have played a pivotal role in understanding the characteristics and behaviour of composite materials, providing valuable insights for their optimization and performance improvement. Techniques like SEM, AFM, XPS, Contact Angle Measurement, and FTIR have enabled researchers to analyse the microstructural features, surface properties, chemical composition, and interfacial interactions within composites. The integration of advanced fabrication and characterization techniques has brought forth numerous benefits for the future of composites. By leveraging these techniques, researchers and engineers can push the boundaries of material design, enabling the development of multifunctional composites with tailored properties and performance characteristics. These advanced composites have the potential to revolutionize various industries, including aerospace, automotive, energy, and healthcare, by offering lightweight structures, improved strength-to-weight ratios, enhanced durability, and multi-functionality [58–60]. Several kinds of research have been conducted for parametric optimization of various manufacturing processes utilizing its significance for multiple applications [61–67].

Furthermore, using advanced techniques to fabricate and characterize composites contributes to the broader sustainability and recycling goals. Many advanced sustainable welding processes, such as Electron beam welding, utilize composite material for welding purposes for various applications [68, 69]. As the global focus on environmental consciousness grows, developing composites with sustainable and recyclable materials becomes crucial. Advanced techniques allow for incorporating recycled materials, bio-inspired designs, and environmentally friendly processes, aligning with the principles of circular economy and reducing the environmental impact of composites [70].

Hence, based on the above analysis of the art, the future directions in this context are listed below.

1. Multifunctional Composites: One of the primary future directions in composite materials is the integration of multi-functionality. Multifunctional composites possess the ability to exhibit a combination of mechanical, electrical, thermal, and other desired properties. This integration opens opportunities for diverse aerospace, automotive, energy, and healthcare applications. Future efforts should focus on developing advanced manufacturing techniques, optimization of material combinations, and characterization methods to realize the full potential of multifunctional composites.

2. Recycling and Sustainability: As sustainability becomes an increasingly important aspect of material design, the future of composite materials lies in developing environmentally friendly manufacturing processes and enhancing recycling capabilities. Composite recycling poses unique challenges due to the complexity of material composition. Future directions should prioritize the development of efficient and cost-effective recycling methods, including advanced sorting techniques, composite disassembly technologies, and innovative recycling strategies. Additionally, incorporating sustainable raw materials and bio-based resins into composite production can enhance their eco-friendliness.

3. Nanocomposites: Nanotechnology offers immense potential for enhancing composite materials' mechanical, electrical, and thermal properties. Nanocomposites, which involve the incorporation of nanoparticles or nanofillers into a matrix, have shown remarkable improvements in strength, stiffness, and functionality. Future directions in nanocomposites should focus on exploring new nanomaterials, developing scalable synthesis methods, and understanding the structure–property relationships at the nanoscale. Integrating nanotechnology into composite materials opens avenues for lightweight and high-performance applications across various industries.

4. Bioinspired Composites: Nature has provided a wealth of inspiration for developing advanced composite materials. Bioinspired composites mimic the hierarchical structures, self-healing properties, and other unique characteristics found in natural materials. Future directions should involve a deeper understanding of the underlying principles governing natural materials and translating these principles into synthetic composites. Combining bioinspired design strategies, such as self-healing mechanisms, multiscale architectures, and adaptive functionalities, composite materials can improve performance, durability, and sustainability.

In conclusion, the future of composite materials relies on embracing multi-functionality, promoting recycling and sustainability, harnessing nanotechnology, and drawing inspiration from bioinspired design. Advancements in advanced fabrication and characterization techniques have revolutionized the field, expanding possibilities for high-performance structures and deepening our understanding of composites. Continued development and implementation of these techniques will shape the future of composites, enabling multifunctional applications and sustainable practices across industries. Research, collaboration, and innovation in these areas will drive the evolution of advanced composite materials.

REFERENCES

[1] A. Babbar, A. Sharma, V. Jain and A.K. Jain, Rotary ultrasonic milling of C/SiC composites fabricated using chemical vapor infiltration and needling technique, *Materials Research Express*, 6, 2019, p. 085607.

[2] R.R. Navagally, Composite materials – History, types, fabrication techniques, advantages, and applications, *International Journal of Mechanical and Production Engineering*, 5, 2017, pp. 82–87.

[3] A. Mohata, N. Mukhopadhyay and V. Kumar, CRITIC-COPRAS-Based Selection of Commercially Viable Alternative Fuel Passenger Vehicle, in Advances in Modelling and Optimization of Manufacturing and Industrial Systems: Select Proceedings of CIMS 2021 2023 Feb 24 (pp. 51–69). Singapore: Springer Nature Singapore CRITIC-COPRAS-based selection of commercially viable alternative fuel, 2023.

[4] Ammar H. Elsheikh, Hitesh Panchal, S. Shanmugan, T. Muthuramalingam, Ahmed M. El-Kassas, and B. Ramesh, Recent progresses in wood-plastic composites: Pre-processing treatments, manufacturing techniques, recyclability and eco-friendly assessment, *Cleaner Engineering and Technology*, 8, 2022, p. 100450.

[5] M.R. Sanjay, S. Siengchin, J. Parameswaranpillai, M. Jawaid, C.I. Pruncu, A. Khan, A comprehensive review of techniques for natural fibers as reinforcement in composites: Preparation, processing and characterization. *Carbohydrate Polymers*. 2019 Mar 1;207:108–121.

[6] S.N. Jamaludin, F. Mustapha, D.M. Nuruzzaman, S.N. Basri, A review on the fabrication techniques of functionally graded ceramic-metallic materials in advanced composites. *Scientific Research and Essays*, 2013 Jun 4;8(21):828–840

[7] S. Nur, S. Jamaludin, F. Mustapha, D.M. Nuruzzaman, Siti Nur Sakinah Jamaludin, Faizal Mustapha et al., A review on the fabrication techniques of functionally graded ceramic-metallic materials in advanced composites, *Academic Journals*, 8, 2013, pp. 828–840.

[8] V. Kumar, C. Prakash, A. Babbar, S. Choudhary, A. Sharma, and A.S. Uppal, Additive manufacturing in biomedical engineering: present and future applications, in *Additive Manufacturing Processes in Biomedical Engineering*. CRC Press, 2022, pp. 143–164.

[9] V. Kumar, A. Babbar, A. Sharma, R. Kumar, and A. Tyagi, Polymer 3D bioprinting for bionics and tissue engineering applications, in *Additive Manufacturing of Polymers for Tissue Engineering*, CRC Press, Boca Raton, 2022, pp. 17–39.

[10] A. Babbar, V. Jain, D. Gupta, A. Sharma, C. Prakash, V. Kumar et al., Additive manufacturing for the development of biological implants, scaffolds, and prosthetics, In *Additive Manufacturing Processes in Biomedical Engineering*, 2022, pp. 27–46, CRC Press.

[11] A. Babbar, V. Jain, D. Gupta, C. Prakash, S. Singh, and A. Sharma, 3D bioprinting in pharmaceuticals, medicine, and tissue engineering applications, in *Advanced Manufacturing and Processing Technology*, 2020, pp. 147–161, Elsevier, Netherland.

[12] A. Babbar, A. Sharma, S. Bansal, J. Mago, and V. Toor, Potential applications of three-dimensional printing for anatomical simulations and surgical planning, in *Materials Today Proceedings*, 33, 2019, pp. 1558–1561.

[13] A. Babbar, Y. Tian, V. Kumar, and A. Sharma, 3D bioprinting in biomedical applications, in *Additive Manufacturing of Polymers for Tissue Engineering*, CRC Press, Boca Raton, 2022, pp. 1–16.

[14] E. Oromiehie, B.G. Prusty, P. Compston, and G. Rajan, Automated fibre placement based composite structures: Review on the defects, impacts and inspections techniques. *Composite Structures*, 224, 2019 Sep 15, p. 110987.

[15] A. Brasington, C. Sacco, J. Halbritter, R. Wehbe, and R. Harik. Automated fiber placement: A review of history, current technologies, and future paths forward. *Composites Part C: Open Access*, 16, 2021 Oct 1, p. 100182.

[16] Z. August, G. Ostrander, J. Michasiow, and D. Hauber, Recent developments in automated fiber placement of thermoplastic composites, *SAMPE Journal*, 50, 2014, pp. 30–37.

[17] C. Red, The outlook for thermoplastics in aerospace composites, 2014–2023, *High-Performance Composites*, 22, 2014, pp. 54–63.

[18] K. Bilisik, Three-dimensional braiding for composites: A review, *Textile Research Journal*, 83, 2013, pp. 1414–1436.

[19] Y. Chai, Y. Wang, Z. Yousaf, M. Storm, N.T. Vo, K. Wanelik et al., Following the effect of braid architecture on performance and damage of carbon fibre/epoxy composite tubes during torsional straining, *Composites Science and Technology*, 10, 2020, pp. 108451.

[20] X. Gao, Q. Wang, B. Sun, B. Gu and M. Hu, Braiding procedure and axial compressive behaviors of three-dimensional integrated braided composite three-way circular tubes, *Textile Research Journal*, 22, 2023, 00405175231158114.

[21] Y.D. Boon, S.C. Joshi, and S.K. Bhudolia, Filament winding and automated fiber placement with in situ consolidation for fiber reinforced thermoplastic polymer composites. *Polymers*, 13, 2021 Jun 11, p. 1951.

[22] M. Azeem, H.H. Ya, M.A. Alam, M. Kumar, P. Stabla, M. Smolnicki, L. Gemi, R. Khan, T. Ahmed, Q. Ma, and M.R. Sadique, Application of filament winding technology in composite pressure vessels and challenges: a review. *Journal of Energy Storage*, 49, 2022, 103468.

[23] N. Minsch, F.H. Herrmann, T. Gereke, A. Nocke, and C. Cherif, Analysis of filament winding processes and potential equipment technologies, *Procedia CIRP*, 66, 2017, pp. 125–130.

[24] Md.M. Billah, Md.S. Rabbi, and A. Hasan, A review on developments in manufacturing process and mechanical properties of natural fiber composites, *Journal of Engineering Advancements*, 2, 2021, pp. 13–23.

[25] G. Gamboa, S. Mazumder, N. Hnatchuk, J.A. Catalan, D. Cortes, I.K. Chen et al., 3D-printed and injection molded polymer matrix composites with 2D layered materials, *Journal of Vacuum Science & Technology A*, 38, 2020, pp. 042201.

[26] S.-J. Liu, Injection molding in polymer matrix composites, in *Manufacturing Techniques for Polymer Matrix Composites (PMCs)*, 2012, pp. 15–46, Woodhead Publishing.

[27] J. Wang, Q. Mao, N. Jiang, and J. Chen, Effects of injection molding parameters on properties of insert-injection molded polypropylene single-polymer composites, *Polymers*, 13, 2021, p. 180–1676.

[28] K. Minchenkov, A. Vedernikov, A. Safonov, and I. Akhatov, Thermoplastic pultrusion: A review. *Polymers*, 13, 2021, p. 180.

[29] P. Esfandiari, J.F. Silva, P.J. Novo, J.P. Nunes, and A.T. Marques, Production and processing of pre-impregnated thermoplastic tapes by pultrusion and compression moulding, *Journal of Composite Materials*, 56, 2022, pp. 1667–1676.

[30] U. Riedel, 10.18 - Biocomposites: Long natural fiber-reinforced biopolymers, in *Polymer Science: A Comprehensive Reference: Volume 1–10*, 2012, pp. 295–315.

[31] E. Barkanov, P. Akishin, and E. Namsone-Sile, Effectiveness and productivity improvement of conventional pultrusion processes, *Polymers*, 14, 2022, p. 841.

[32] J.J. Murray, C. Robert, K. Gleich, E.D. McCarthy, and C.M. Ó. Brádaigh, Manufacturing of unidirectional stitched glass fabric reinforced polyamide 6 by thermoplastic resin transfer moulding, *Materials and Design*, 189, 2020, 108512.

[33] N. Razali, M.R. Mansor, G. Omar, S.A.F.S. Kamarulzaman, M.H. Zin, and N. Razali, Out-of-autoclave as a sustainable composites manufacturing process for aerospace applications, in *Design for Sustainability: Green Materials and Processes*, 2021, pp. 395–413, Elsevier.

[34] B. Miranda Campos, S. Bourbigot, G. Fontaine, and F. Bonnet. Thermoplastic matrix-based composites produced by resin transfer molding: A review. *Polymer Composites*, 43, 2022, pp. 2485–2506.

[35] B. Gartner, V. Yadama, and L. Smith, Resin transfer molding of wood strand composite panels, *Forests*, 13, 2022, p. 278.

[36] A.P. Acosta, A.A. Xavier da Silva, R. de Avila Delucis, and S.C. Amico, Wood and wood-jute laminates manufactured by vacuum infusion, *Journal of Building Engineering*, 64, 2023, 105619.

[37] K.K. Verma, B.L. Dinesh, K. Singh, K.M. Gaddikeri, and R. Sundaram, Challenges in processing of a cocured wing test box using vacuum enhanced resin infusion technology (VERITy), *Procedia Materials Science*, 6, 2014, pp. 331–340.

[38] V. Kumar, P.P. Das, and S. Chakraborty, Grey-fuzzy method-based parametric analysis of abrasive water jet machining on GFRP composites, *Sādhanā*, 45, 2020, p. 106.

[39] V. Kumar, S. Diyaley, and S. Chakraborty, Teaching-learning-based parametric optimization of an electrical discharge machining process, *Facta Universitatis, Series: Mechanical Engineering*, 18, 2020, pp. 281–300.

[40] A. Sharma, V. Kumar, A. Babbar, V. Dhawan, K. Kotecha, and C. Prakash, Experimental investigation and optimization of electric discharge machining process parameters using grey-fuzzy-based hybrid techniques, *Materials*, 14, 2021, p. 5820.

[41] C. Prakash, V. Kumar, A. Mistri, A.S. Uppal, A. Babbar, B.P. Pathri et al., Investigation of functionally graded adherents on failure of socket joint of FRP composite tubes, *Materials*, 14, 2021, p. 6365.

[42] A. Babbar, V. Jain, D. Gupta, and D. Agrawal, Histological evaluation of thermal damage to osteocytes: A comparative study of conventional and ultrasonic-assisted bone grinding, *Medical Engineering & Physics*, 90, 2021, pp. 1–8.

[43] L.M. Khaskhanova, S.N. Razumova, D.M. Serebrov, Z.A. Gureva, A.V. Vetchinkin, A.V. Rebrii et al., Scanning electron microscopy, *Journal of International Dental and Medical Research*, 15, 2022, pp. 107–110.

[44] A. Kumar, A. Babbar, V. Jain, D. Gupta, B.P. Pathri, C. Prakash et al., Investigation and enhancement of mechanical properties of SS-316 weldment using TiO2-SiO2-Al2O3 hybrid flux, *International Journal on Interactive Design and Manufacturing*, 2023, 1–7.

[45] F.J. Giessibl, Advances in atomic force microscopy. *Reviews of Modern Physics*, 75, 2003, p. 949.

[46] A. Yadav, P. Rohru, A. Babbar, R. Kumar, N. Ranjan, J.S. Chohan et al., Fused filament fabrication: A state-of-the-art review of the technology, materials, properties and defects. *International Journal on Interactive Design and Manufacturing*, 2022, 1–23.

[47] A. Babbar, V. Jain, D. Gupta, S. Singh, C. Prakash and C. Pruncu, Biomaterials and fabrication methods of scaffolds for tissue engineering applications, *3D Printing in Biomedical Engineering*, 2020, pp. 167–186.

[48] M. Scopelliti, X-ray photoelectron spectroscopy, in *Spectroscopy for Materials Characterization*, 2021, pp. 351–382.

[49] A. Sharma, V. Jain, D. Gupta and A. Babbar, A review study on miniaturization: A boon or curse, *Advanced Manufacturing and Processing Technology*, 2020, pp. 111–131.

[50] A. Babbar, V. Jain, D. Gupta, D. Agrawal, C. Prakash, S. Singh et al., Experimental analysis of wear and multi-shape burr loading during neurosurgical bone grinding, *Journal of Materials Research and Technology*, 12, 2021, pp. 15–28.

[51] A. Babbar, V. Jain, and D. Gupta, In vivo evaluation of machining forces, torque, and bone quality during skull bone grinding, *Proceedings of the Institution of Mechanical Engineers. Part H*, 234, 2020, pp. 626–638.

[52] S.L. Schellbach, S.N. Monteiro, and J.W. Drelich, *A novel method for contact angle measurements on natural fibers*, *Materials Letters*, 164, 2016, pp. 599–604.

[53] A. Sharma, A. Babbar, V. Jain, and D. Gupta, Enhancement of surface roughness for brittle material during rotary ultrasonic machining, *MATEC Web of Conferences*, 249, 2018, p. 01006.

[54] D. Singh, A. Babbar, V. Jain, D. Gupta, S. Saxena, and V. Dwibedi, Synthesis, characterization, and bioactivity investigation of biomimetic biodegradable PLA scaffold fabricated by fused filament fabrication process, *Journal of the Brazilian Society of Mechanical Sciences and Engineering*, 41, 2019.

[55] A. Babbar, V. Jain, D. Gupta, and D. Agrawal, Finite element simulation and integration of CEM43°C and Arrhenius Models for ultrasonic-assisted skull bone grinding: A thermal dose model, *Medical Engineering & Physics*, 90, 2021, pp. 9–22.

[56] A. Sharma, V. Grover, A. Babbar, and R. Rani, A trending nonconventional hybrid finishing/machining process, in *Non-Conventional Hybrid Machining Processes*, First edition, CRC Press, Boca Raton, 2020, pp. 79–93.

[57] R. Wang, and Y. Wang, Fourier transform infrared spectroscopy in oral cancer diagnosis. *International Journal of Molecular Sciences*, 22, 2021, p. 1206.

[58] S. Das, B. Sarkar, and V. Kumar, RIM-based performance evaluation of DLC coating under conflicting environment, in Advances in Modelling and Optimization of Manufacturing and Industrial Systems: Select Proceedings of CIMS 2021, 2023, Feb 24, pp. 303–320, Singapore: Springer Nature Singapore.

[59] V. Kumar, K. Kalita, P. Chatterjee, E.K. Zavadskas, and S. Chakraborty, A SWARA-CoCoSo-Based Approach for Spray Painting Robot Selection, *Informatica*, 33, 2021, pp. 35–54.

[60] S. Chakraborty, and V. Kumar, Development of an intelligent decision model for non-traditional machining processes, *Decision Making: Applications in Management and Engineering*, 4, 2021, pp. 194–214.

[61] S. Chakraborty, P.P. Das, and V. Kumar, Application of grey-fuzzy logic technique for parametric optimization of non-traditional machining processes, *Grey Systems Theory and Application*, 8, 2018, pp. 46–68.

[62] V. Kumar, and S. Chakraborty, Analysis of the surface roughness characteristics of EDMed components using GRA method, In Proceedings of the International Conference on Industrial and Manufacturing Systems (CIMS-2020) Optimization in Industrial and Manufacturing Systems and Applications 2022, 2022, pp. 461–478, Springer International Publishing.

[63] S. Chakraborty, P.P. Das, and V. Kumar, A Grey fuzzy logic approach for cotton fibre selection, *Journal of The Institution of Engineers (India): Series E*, 98, 2017, pp. 1–9.

[64] A. Babbar, C. Prakash, S. Singh, M.K. Gupta, M. Mia, and C.I. Pruncu, Application of hybrid nature-inspired algorithm: Single and bi-objective constrained optimization of magnetic abrasive finishing process parameters, *Journal of Materials Research and Technology*, 9, 2020, pp. 7961–7974.

[65] A. Babbar, V. Jain, D. Gupta, and C. Prakash, Experimental investigation and parametric optimization of neurosurgical bone grinding under bio-mimic environment, *Surface Review and Letters*, 30, 2023, 2141005.

[66] A. Babbar, A. Sharma, and M. Chugh, Application of flexible sintered magnetic abrasive brush for finishing of brass plate, *Optimization in Engineering Research*, 01, 2020, pp. 36–47.

[67] S. Chakraborty, V. Kumar, and K. Ramakrishnan, Selection of the all-time best World XI Test cricket team using the TOPSIS method, *Decision Science Letters*, 8, 2018, pp. 95–108.

[68] A. Kundu, D.K. Pratihar, D. Chakrabarti, and V. Kumar, Introduction to the new emerging micro-electron beam welding technology: A sustainable manufacturing, in *Futuristic Manufacturing*, 2023, pp. 127–141, CRC Press.

[69] M. Kumar, A. Babbar, A. Sharma, and A.S. Shahi, Effect of post weld thermal aging (PWTA) sensitization on micro-hardness and corrosion behavior of AISI 304 weld joints, *Journal of Physics: Conference Series*, 1240, 2019, 012078.

[70] A. Sharma, M. Kalsia, A.S. Uppal, A. Babbar, and V. Dhawan, Machining of hard and brittle materials: A comprehensive review, *Materials Today Proceedings*, 50, 2021, pp. 1048–1052.

2 Machining of Aluminium Silicon Carbide Metal Matrix Composite Using Fabricated Electro Stream Drilling Setup

Manpreet Singh
Chitkara University Institute of Engineering and Technology, Rajpura, India

2.1 INTRODUCTION

Various advanced manufacturing processes are developed to improve the functional performance of the machines and processes [1–4]. In micro-manufacturing, the electrochemical machining processes are playing an important role in the industries [5, 6]. These processes are capable to manufacture the complex shapes at micro-level with a fine finished surface. The surface roughness and other characteristics of the manufactured components depend upon the electrolytic concentration, pressure of the electrolyte and DC power supply [7]. In almost every electrochemical machining process, these parameters play a crucial role to maintain the quality of the manufactured components. The micro-machining process is very important in many manufacturing industries, viz. aerospace, bio-medical, electrical, and electronics, automobile, thermal power plants, nuclear power plants, etc., because of the use of miniature product in such industries. Scientists and metallurgist have developed different advanced materials whose detailed properties and applications are available but processing of these materials to make miniature parts are very difficult because of their high strength, high wear resistance, and electrical conductivity [8]. Aerospace and turbine industries require a large range number of ceramics and composite materials because of their high strength and low density. However, these materials have poor machinability which is one of the barriers to resist their application in modern advanced industries. The machinability issue with composites cannot be solved by well-known conventional or unconventional machining procedures [9]. Electro stream drilling (ESD) is an effective advanced drilling process to create macro and micro holes. The miniaturization of the industrial components is an important aspect to lower the production cost of high-volume products, reducing the power consumption and thus improving the efficiency of the process. Therefore, electrochemical micro-machining (ECMM) finds its application in machining of composites. The ECMM setup comprises various systems and subcomponents. These include the mechanically controlled devices, electrically controlled devices and hydraulic control system. ECMM is performed in the small inter electrode gap to achieve the desired accuracies. It has been observed that the miniaturization of the narrow gaps increases the process complexities. The accuracy and capability of the machining are improved by using a tool with a tiny helical shape [10]. When the acidic electrolyte is passed through the nozzle it gets converted into the negatively charged acidic electrolyte. This negatively charged acidic electrolyte stream is intruded on the surface of the workpiece by using a finely drawn brass nozzle [11]. A pressure pump is provided to generate the required pressure [12]. The acidic electrolyte is transferred under the pressure of 3–10 bars through the nozzle using the pressure pump [13]. The acidic electrolyte stream behaves like a cathode and the workpiece as an anode [13–17]. An electric potential is developed between the nozzle stream and the workpiece surface [18, 19]. The material gets eroded in the form of ions due to the impingement of acidic electrolyte stream on the surface of workpiece [20]. These ions are further carried away by the acidic electrolyte flow [21]. The acidic electrolyte stream is made to follow the lean flow path to improve the process performance [22–26]. For maintaining the current flow, a sufficient DC power supply is required (150–850 V) which provides the electrical potential between the nozzle and the workpiece surface [23]. In recent times, various studies have been performed to study various machining operations on different kinds of material [27–54]. In the present work, the ESD setup is fabricated. The setup is utilized for drilling the aluminium/silicon carbide metal matrix composite (Al/SiC-MMC) workpieces. The workpiece metal matrix composite is casted by using melt-stir technique. Al/SiC-MMC material has various industrial applications where miniaturization machining is required. Based on the industrial applications, the Al/SiC-MMC is cast. The Taguchi method is utilized for the optimization of the process parameters which affects the material removal rate (MRR) of the ESD process.

2.2   FABRICATION OF ELECTRO STREAM DRILLING SETUP

The fabricated setup is shown in Figure 2.1. Each unit has its own importance in the setup. At first, the base is fabricated in such a manner that the entire units of the setup are placed at the same level. The glass chamber is fabricated using the five glass pieces. These pieces are attached together to create the workspace space inside the glass chamber which is depicted in Figure 2.2(a). The high-pressure pump is utilized in order to circulate the electrolyte. This pump is operated with the help of 0.5 hp AC motor as shown in Figure 2.2(b). The X-axis nozzle holding vise is fabricated so as to hold the different flat strips. The X slider is used to hold the nozzle which is a part of the Vernier caliper. To hold the workpiece, the mild steel plate holder is fabricated.

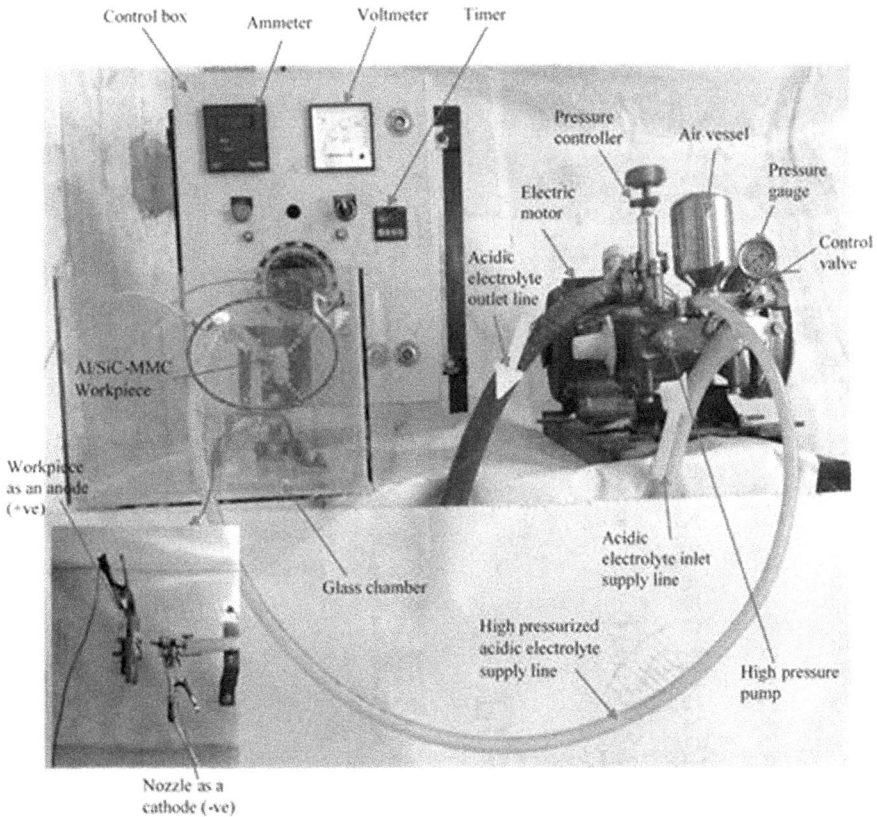

FIGURE 2.1 Fabricated experimental setup of the electro stream drilling process.

This holder can easily accommodate different sizes of workpieces. A number of holes are drilled and with the help of nut and bolt the workpiece is clamped. To obtain variable DC supply, the main AC supply of 220 volts is regulated.

The DC supply is varied from 170 to 260 volts by using a stepper up transformer with a regulated machining setup. As seen in Figure 2.2(c), the DC power supply control unit is used to regulate the DC power supply that is used for experiments.

For DC power supply, the anode and the cathode terminals are attached to the copper nozzle and to the workpiece as shown in Figure 2.2(d). The working gap is maintained between the end of the nozzle and the exposed surface of the workpiece. This is done to generate the necessary electrochemical potential for the drilling operation.

The electro jet drilling setup consists of precision machined components that are manually controlled by moving the nozzle up and down vertically. An electrolyte storage arrangement, glass machining chamber, job fixing vice, electric motor, high pressure pump of capacity 0–10 bar, AC motor of 1hp, 1450 rpm and 240 volts are used to operate the high-pressure pump.

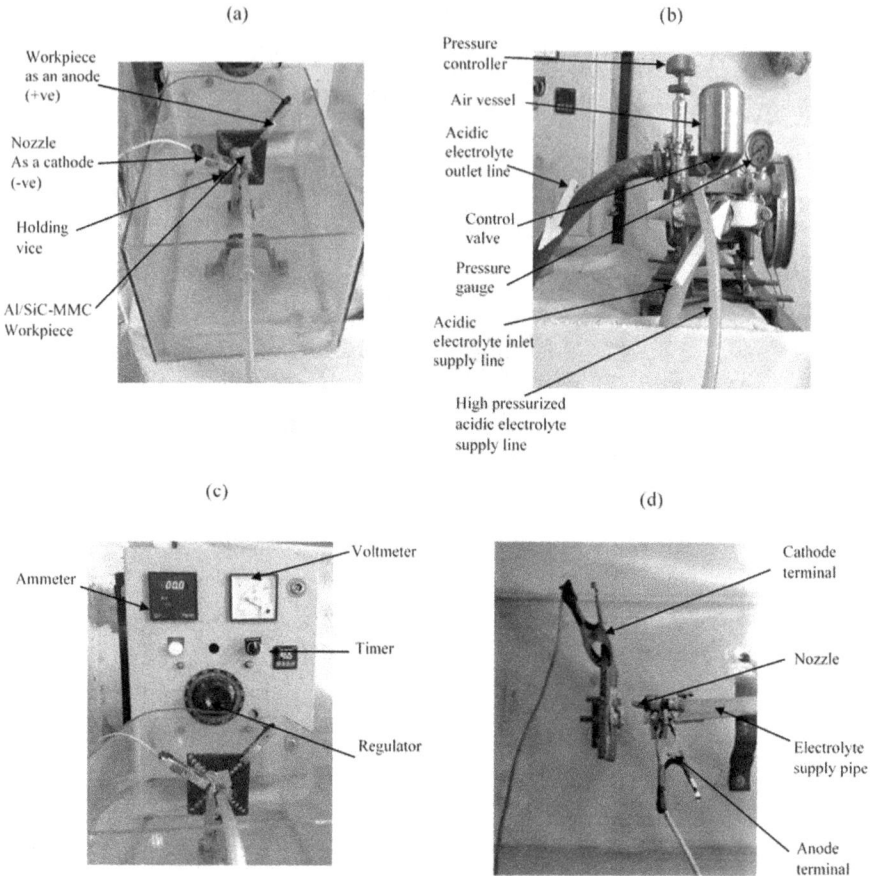

FIGURE 2.2 The units of experimental setup (a) glass chamber with working space, (b) high pressure pump with electric motor, (c) dc power supply controller, and (d) anode cathode terminal with nozzle holder.

2.3 FABRICATION OF WORKPIECE MATERIAL ALUMINIUM SILICON CARBIDE METAL MATRIX COMPOSITE (AL/SIC-MMC)

Silicon Carbide (SiC) reinforced particles with an average size of 200 mesh is considered for casting as shown in Figure 2.3(a). A number of flat plates of different dimensions with 7 vol%, SiC reinforced particles were fabricated and are utilized to manufacture the workpiece samples. Initially, preheating of aluminium alloy was done. The preheating temperature was 450°C for the time period of 2 hours. After this process, SiC particles were preheated at 1100°C for 90 minutes. This is done to enhance the wettability property. The enhancement is achieved by eliminating the absorbed hydroxide and other gases. The furnace temperature was first elevated above the melting point temperature of 710°C. It is done to melt the matrix completely. The matrix was allowed to cool at a temperature just below the melting point

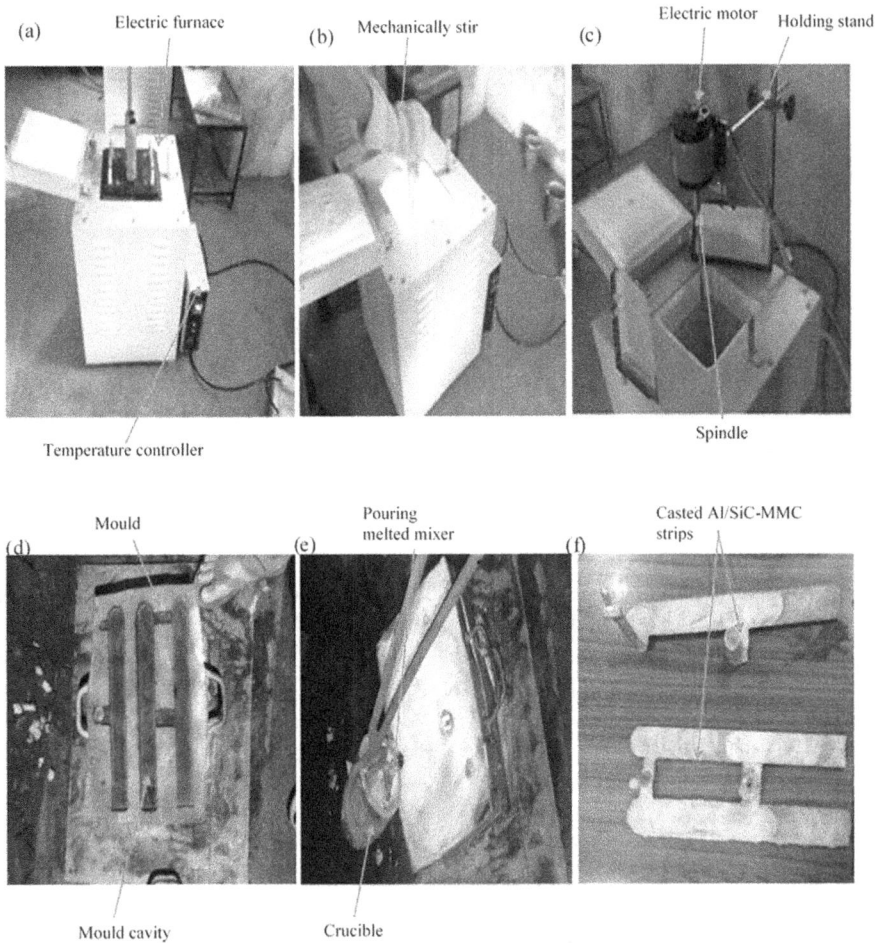

FIGURE 2.3 The processing steps of the melt-stir casting technique are as follows: (a) melt-stir casting setup, (b) churning of mixture mechanically, (c) mixing with stir, (d) impression in the mould with 5 mm thick strip, (e) pouring the melting mixer, and (f) removing the casting.

in order to preserve the slurry in a semi-solid state after melting. The warmed SiC particles were then added and mechanically mixed, as shown in Figure 2.3(b). After that, heat was applied to the composite slurry until it turned liquid. At this stage, the mechanical mixing was done for 20 minutes at a mean speed of 200 rpm as shown in Figure 3(c). At the last stage of the process, the furnace temperature was controlled within $710 \pm 10°C$. Moulds with size 45 mm width and 330 mm long were made using 5 mm thick steel sheets as shown in Figure 2.3(d). The molten Al/SiC-MMC is poured in the mould which is shown in Figure 2.3(e). The sheets were preheated to 350°C for 2 hours before pouring the molten Al/SiC-MMC. The mould is thus broken and the 5 mm thick Al/SiC-MMC strips are collected. The strips are then machined and used as workpieces for the experimentation. The XRD pattern obtained from Al/SiC-MMC manufactured by Vortex method in conjunction with the matrix alloys and

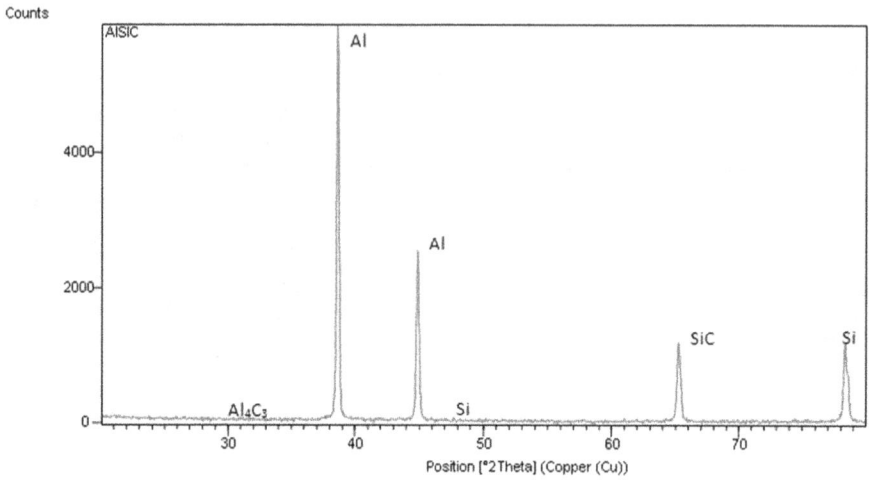

FIGURE 2.4 X-Ray diffraction of Al/SiC-MMC.

the diffraction intensity peaks corresponding to the α-aluminium phase, a solid solution which is rich in aluminium (α-Al) of crystalline FCC structure. The two major contributions which can be identified are a small peak at $2\Theta = 38.82°$ and at $2\Theta = 44.5°$ that corresponds to (111) and (200) reflection of α-Al phase, respectively, as shown in Figure 2.4. On the other hand, the new composite shows two main diffraction intensities: $2\Theta = 65.35°$ and $78.35°$, which correspond to (220) and (311) of SiC particles, respectively, as depicted in Figure 2.4.

Aluminium alloy (AA6063)/SiC composites have excellent thermal conductivity, high shear strength, great abrasion resistance, high temperature operation, inflammability, little reaction with fuels or solvents, and the capacity to be made and processed using non-traditional machinery. The characteristics of the commercially available Al (AA 6063) alloy utilized as Al-matrix are shown in Table 2.1. The density of aluminium is very low, i.e. 2.71 g/cc and melting point is 660°C. It has a tensile strength and shear strength of 100 MPa and 70 MPa, respectively.

TABLE 2.1
Composition of Aluminium Workpiece

Elements of Al 6063	%
Si	0.44
Mn	0.07
Mg	0.6
Cu	0.018
Fe	0.2
Ti	0.008
Cr	0.005
Al	98.65

2.4   EXPERIMENTAL PLANNING AND DESIGN OF EXPERIMENTS

Taguchi method-based robust design L_{16} (4^5) orthogonal arrangement was used for experimental analysis. Material removal rate was determined by using weight difference of the workpiece before and after each micro drilling operation. Electronic balance with resolution 0.001 g was used to weigh the workpieces during pre- and post-operation run. There are a number of studies available using Taguchi's Orthogonal Array (OA) with fewer experimental runs. The objective functions used in the optimization process are Taguchi's signal-to-noise ratios (S/N), which are logarithmic functions of the desired output. Each experiment was performed thrice. The mean values of the responses were recorded. The acquired data was further used for the analysis.

Table 2.2 reports the details of the developed micro drilling setup, brass nozzles, Al/SiC-MMC work-piece and electrolyte used for the experimentation. Table 2.3 reports the process parameters which used during experimentation.

The experiments were performed in the random order. This was done to negate the effect of setup constraints or environmental factors. These factors or constraints were not taken into consideration initially but they could alter the responses.

TABLE 2.2
Detail of Experimental Conditions

Machine Tool Used	Micro Electro Jet Drilling Setup
Electrolyte Concentration	i. 11 g of $NaNo_3$ + 50 ml of H_2So_4/Litre of Water
	ii. 14 g of $NaNo_3$ + 50 ml of H_2So_4/Litre of Water
	iii. 17 g of $NaNo_3$ + 50 ml of H_2So_4/Litre of Water
	iv. 20 g of $NaNo_3$ + 50 ml of H_2So_4/Litre of Water
Workpiece	Al/SiC-MMC
Workpiece thickness	3 mm
Dimensions	40 mm × 40 mm
Tool Used	Brass Nozzles (0.2, 0.3, 0.4, 0.5 mm)

TABLE 2.3
Machining Parameters and its Ranges

Sr. No	Machining Parameters	Ranges				Units
		1	2	3	4	
1	DC supply voltage (A)	170	200	230	260	volt
2	Pump pressure (B)	2	4	6	8	bar
3	Electrolyte concentration (C)	16	19	22	25	% by weight
4	Feed rate (D)	0.20	0.25	0.30	0.35	mm/min
5	Nozzle diameter (E)	0.2	0.3	0.4	0.5	mm

TABLE 2.4

Experimental Trails Combinations with Response (Material Removal Rate) and S/N Ratio Values

Exp.	A	B	C	D	E	MRR (mg/min) Y1	Y2	Y3	Avg. MRR (mg/min)	S/N ratio
1	170	2	16	0.20	0.2	0.740	0.745	0.750	0.745	5.271
2	170	4	19	0.25	0.3	0.808	0.786	0.821	0.805	5.381
3	170	6	22	0.30	0.4	0.834	0.826	0.837	0.832	5.677
4	170	8	25	0.35	0.5	0.842	0.840	0.838	0.840	5.775
5	200	2	19	0.30	0.5	0.844	0.825	0.836	0.834	5.796
6	200	4	16	0.35	0.4	0.846	0.833	0.845	0.835	5.703
7	200	6	25	0.20	0.3	0.855	0.875	0.866	0.865	5.392
8	200	8	22	0.25	0.2	0.898	0.904	0.892	0.898	5.339
9	230	2	22	0.35	0.3	0.840	0.819	0.830	0.829	5.553
10	230	4	25	0.30	0.2	0.877	0.866	0.888	0.878	5.494
11	230	6	16	0.25	0.5	0.855	0.828	0.888	0.857	5.918
12	230	8	19	0.20	0.4	0.923	0.898	0.913	0.912	5.832
13	260	2	25	0.25	0.4	0.905	0.894	0.892	0.898	6.123
14	260	4	22	0.20	0.5	0.984	0.997	0.980	0.987	6.235
15	260	6	19	0.35	0.2	0.991	0.987	0.971	0.983	5.685
16	260	8	16	0.30	0.3	0.943	0.950	0.938	0.944	5.792

For each trial, specified input parameters were stated and through holes were generated during the machining of Al/SiC-MMC. The experimental combinations and its response values are reported in Table 2.4.

By monitoring the weight difference of eight workpieces before and after each micro-drilling operation, the material removal rate (MRR) is calculated. Electronic balance of resolution 0.001 g was used to weigh the work-pieces before and after each run. MRR is calculated by using Eq. (2.1).

$$\text{MRR} = \frac{\left(W_o - W_i\right)}{\text{Machining Time}} \, \text{mg/min} \qquad (2.1)$$

Where, W_o is the initial weight of the workpiece in kg and W_i is the final weight of the workpiece in kg.

2.4.1 S/N RATIO AND PARAMETRIC OPTIMIZATION FOR MATERIAL REMOVAL RATE

The S/N (dB) for material removal is determined and presented in Table 2.4 using the maxim "larger is better." Eq. (2.2) is utilized to define the summary statistics, (dB), of the larger-the-better, i.e. for material removal (MRR, mg/min).

$$\eta = -10 \log_e \left[\frac{1}{n} \sum_{i=1}^{n} \frac{1}{y_i^2} \right] \qquad (2.2)$$

$i = 1, 2\ldots n$, where n is the number of replications of the ith experiment and y_i is the response value or quality features at the ith experiment. is the S/N ratio in dB.

2.4.2 STATISTICAL MODEL FOR MATERIAL REMOVAL RATE (MG/MIN)

The theoretical equation is anticipated to maximize the rate of material removal during the operation using the greater is better approach of the Taguchi method. Using this equation, the final MMR value was predicted and interaction graphs are then plotted. The theoretical equation developed for MRR is stated below:

$$\begin{aligned}
Y_{MRR} = {} & 90.7 \times 10^{-2} - 1.80 \times 10^{-4} A + 1.61 B \times 10^{-2} + 4.80 \times 10^{-2} C \\
& + 80.22 \times 10^{-2} D + 5.78 \times 10^{-5} E + 7.56 \times 10^{-5} AC + 3.05 \times 10^{-3} A.D \\
& + 3.47 \times 10^{-3} BC - 6.06 \times 10^{-2} BE - 4.75 \times 10^{-2} CE + 2.32 \times 10^{-6} A^2 \\
& - 2.83 \times 10^{-4} B^2 - 1.20 \times 10^{-3} C^2 + 61.51 \times 10^{-2} D
\end{aligned} \tag{2.3}$$

Where, A is the dc power supply in volts, B is pump pressure in bar, C is the electrolyte concentration in % by weight, D is the feed rate in mm/min, and E is the nozzle diameter in mm.

2.5 RESULTS AND DISCUSSIONS

The S/N ratio (dB) graph for MRR is shown in Figure 2.5. It is clear that a DC power supply of 260 volts, 4 bar of pump pressure, 25% by weight of the electrolyte concentration, reciprocation motion of 0.25 mm/min, and nozzle diameter of 0.5 mm are

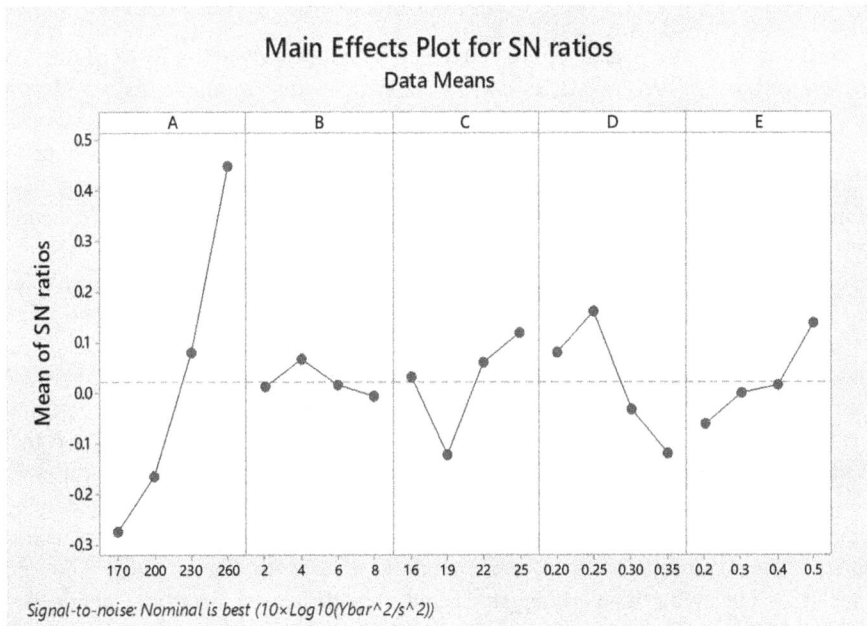

FIGURE 2.5   Plot between the S/N ratio (dB) values and machining parameters.

TABLE 2.5

Analysis of Variance for the Material Removal Rate

Source	Sum of Squares	DOF	Mean Sum of Squares	F Value	P Value	Contribution (%)
A	0.619	1	0.619	264.14	<0.0001	33.29
B	0.024	1	0.024	10.94	0.0042	1.29
C	0.056	1	0.056	23.94	0.0003	3.01
D	0.011	1	0.011	4.99	0.0597	0.59
E	0.467	1	0.467	199.35	<0.0001	25.12
AB	0.021	1	0.021	7.87	0.0052	1.12
AC	0.034	1	0.034	12.87	0.0031	1.82
AE	0.045	1	0.045	15.23	0.0023	2.42
BC	0.053	1	0.053	19.88	0.0003	2.85
BD	0.112	1	0.112	89.89	<0.0001	6.02
CE	0.089	1	0.089	45.67	0.0004	4.78
A2	0.234	1	0.234	132.43	<0.0001	12.58
B2	0.023	1	0.023	9.87	0.0039	1.23
C2	0.034	1	0.034	12.56	0.0033	1.82
D2	0.012	1	0.012	5.23	0.0633	0.64
Error	0.025	32	0.000781			
Total	1.859	47				

the optimal parametric combinations for greater arithmetic mean values of material removal rate (MRR).

The nozzle diameter and supply voltage both significantly affect MRR. Table 2.5 represents the ANOVA for MRR. Here, A, C, E are significant model terms. Also, it is clear from Table 2.5 that the 'F' values against parameters like supply voltage, pump pressure, electrolyte concentration and nozzle diameter reflect higher value in comparison to the other parameters. Hence, the effect of supply voltage is more significant in comparison to the effect of other parameters considered for experimental investigation.

The inverted type Metallurgical Microscope is used to examine the various holes in the workpieces of the aluminium silicon carbide metal matrix composite. The microscopic photographs were taken with the help of microscope.

Figure 2.6(a) depicts the MRR increases as pump pressure increases, while MRR decreases as pump pressure increases in the higher range. There are two potential causes for the decrease in MRR at greater pump pressure. The first explanation is that a faster stream flow rate at higher pump pressure would result in less time for electrolyte stream interaction with the workpiece. The second factor may be linked to the working gap's increased electrolyte stream flow velocity. This causes the temperature of the electrolyte to drop, which further reduces its conductivity. Reduced machining current in the working gap is caused by decreased electrolyte conductivity. The MRR declines as a result of this. As illustrated in Figure 2.6(a), the material

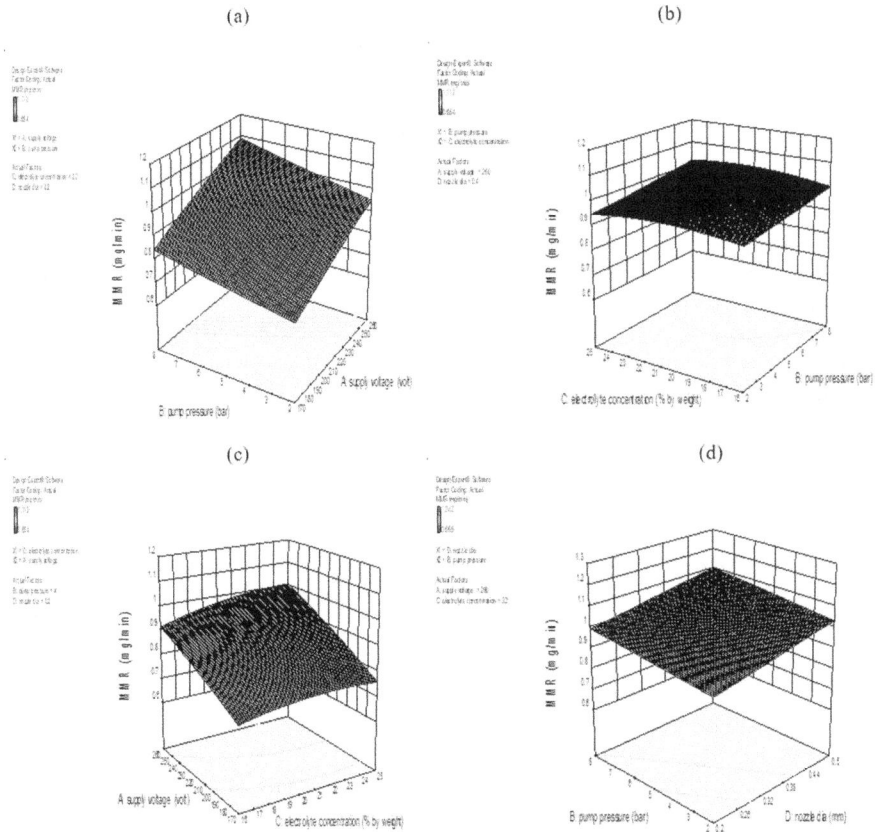

FIGURE 2.6 Material removal rate (MRR) with respect to (a) interaction effect of the supply voltage (A) and the pump pressure (B), (b) interaction effect of the pump pressure (B) and the electrolyte concentration (C) (c) interaction effect of the power supply (A) and the electrolyte concentration (C), and (d) interaction effect of the pump pressure (A) and the nozzle diameter (D).

removal rate (MRR) is improved at lower pump pressure due to an increase in the interaction period between the workpiece and the electrolyte stream.

The association between material removal rate and electrolyte content at various pump pressure levels is depicted in Figure 2.6(b). Higher electrolyte content is reported to accelerate the rate of material clearance. The higher electrical conductivity of the electrolyte is to blame for this. As a result, the working gap's machining current value increases. Additionally, at higher concentrations, a significant number of ions become involved in the machining process, increasing the machining current and ultimately causing an increase in MRR.

The relationship between the rate of material removal and power supply at various electrolyte concentrations is shown in Figure 2.6(c). The machining current would grow across the working gap with any increase in applied voltage. The rate of material removal rises as a result. It can be seen in Figure 2.6(c) that the MRR value increases as the supply voltage and electrolyte concentration increase. This is as a result of the electrolyte stream's higher electrical conductivity with increasing electrolyte concentration. As a result, the working gap's greater voltage potential is maintained.

The nozzle diameter is regarded as an independent parameter when analysing how the electrolyte stream affects the production of the ESD process. The impact of nozzle diameter on MRR at various pump pressure levels is depicted in Figure 2.6(d). It could be observed that the decline in the MRR results when nozzle diameter range is smaller i.e. in between 0.20 and 0.30 mm. The stock removal rate though continues to increase when nozzle diameter is in the higher range which is 0.40–0.50 mm. The enlarged electrolyte flow-path column is because of the increase in the size of the electrolyte stream which eventually is the major reason for the rise in material removal rate. Also, at the higher value of the nozzle diameter the voltage density is increased. Higher the voltage density, higher is the material removal rate. This is also one of the reason that at 0.5 mm diameter of nozzle, the higher material removal rate is observed.

The % error value varies from −5.81 to 6.97 between the experimental and developed mathematical model responses. Table 2.6 represents the parametric condition that are used for additivity test to validate the developed mathematical model of material removal rate (mg/min) during micro drilling of aluminium/silicon carbide metal matrix composite by fabricated ESD setup. Also, the % error in the calculated values by using the current model with respect to the experimental results is also reported.

The actual workpieces of Al/SiC-MMC post experimentation are shown in Figure 2.7. After experimentation, the micro-holes of 0.2 mm, 0.3 mm, 0.4 mm, and 0.5 mm are depicted in Figure 2.7. From Figure 2.7 it is clear that, two sets of trials are performed on each workpiece. The microscopy of the electro drilled holes are done after the experimental trials. These microscopic images are shown in Figure 2.8. The microscope images show the machined holes after stream drilling with fabricated setup. This fabricated setup is very useful for the various industrial applications like in medical, automobile, aerospace, and micro fabrication which further finds its use in electronic and computer components.

TABLE 2.6

Additivity Test to Validate the Developed Mathematical Model of Material Removal Rate (MRR) During Micro Drilling

Exp. No	EJD Setup Parameters				Material Removal Rate (mg/min)		%age Error
	A	**B**	**C**	**D**	**Experimental Values**	**Theoretical Values**	**Error**
1	180	1.5	15	0.22	1.62	1.74	6.97
2	210	3	18	0.27	1.82	1.72	−5.81
3	240	4.5	21	0.32	1.78	1.80	1.11

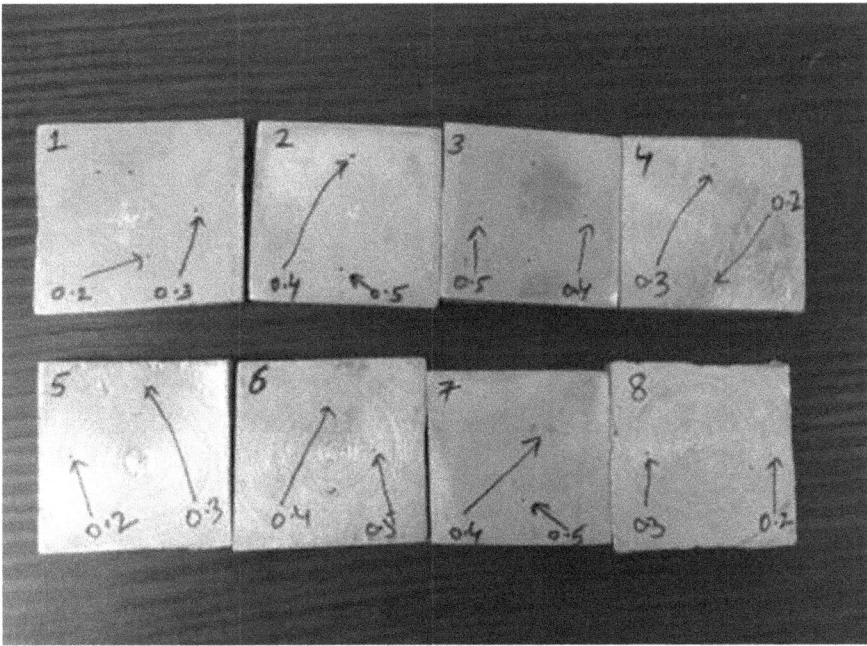

FIGURE 2.7   Actual Al/SiC-MMC workpieces after different experimental trails.

(a)   (b)

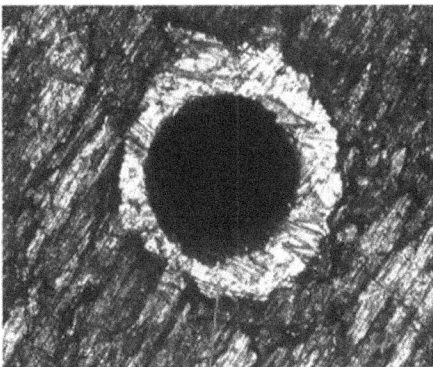

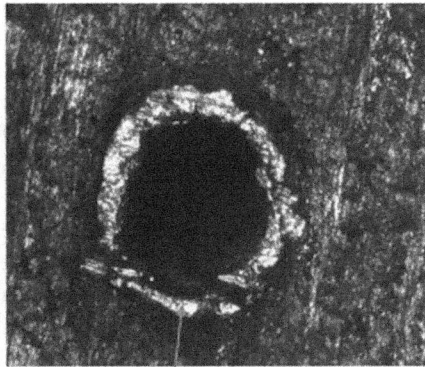

0.5 mm electro   0.4 mm electro
stream drill hole   stream drill hole

FIGURE 2.8   Microscopy of different electro drilled holes at 500 X (a) 0.5 mm drill hole, (b) 0.4 mm drill hole

(*Continued*)

(c)   (d)

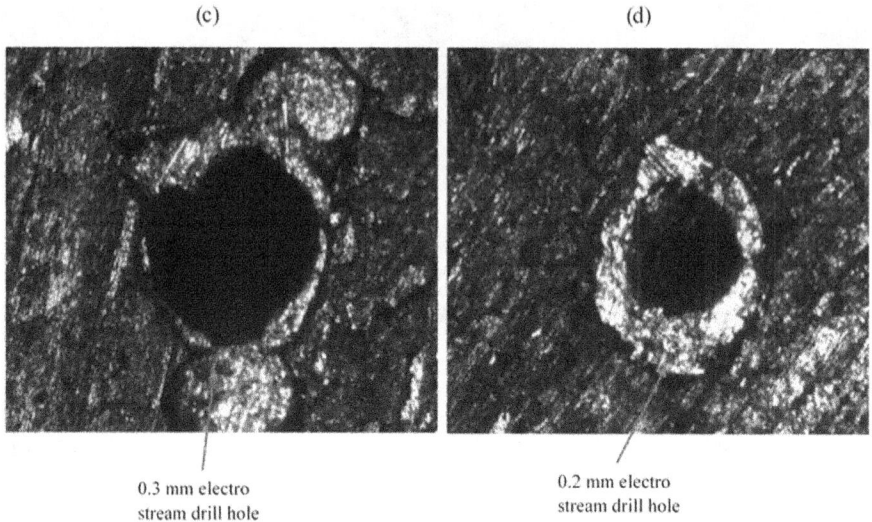

0.3 mm electro   0.2 mm electro
stream drill hole   stream drill hole

FIGURE 2.8 (Continued) Microscopy of different electro drilled holes at 500 X (c) 0.3 mm drill hole, and (d) 0.2 mm drill hole.

2.6 CONCLUSION

On the basis of the experiments results during electro jet drilling of electrically conductive high strength, wear resistance Al/SiC-MMC on developed ESD setup and thereafter discussion on the investigated results the following points are concluded listed below.

- In the developed setup of ESD, parameters such as DC supply voltage and nozzle diameter are considered to be the most significant parameters as their contribution in material removal rate are 33.29% and 25.12%, respectively.
- The parametric combination for achieving the highest material removal rate is at 260 volts DC power supply, 4 bar pump pressure, 25% electrolyte concentration by weight, 0.25 mm/min of feed rate, and nozzle diameter of 0.5 mm.
- The mathematical model for material removal rate is effectively proposed for estimation of parametric value in advance. This process is useful for various industrial applications where minuscule metal parts are fabricated. The optimized parameters are useful in controlling the capability of the ESD process in various industrial applications.
- Uniform feed is an important aspect with uniform pump pressure and supply voltage for effective machining on developed electro jet drilling setup.

REFERENCES

1. Ahn SH, Ryu SH, Choi DK, Chu CN. Electrochemical Micro Drilling Using Ultra Short Pulses. *Precis Eng* 28(2), 129–34 (2004).
2. Hai-Ping Tsui HP, Hung JC, You JC, Yan BH. Improvement of Electrochemical Micro Drilling Accuracy Using Helical Tool. *Mater Manuf Process* 23(5), 499–505 (2008).

3. Bhattacharyya B, Doloi B, Sridhar PS. Electrochemical Micro-Machining: New Possibilities for Micro-Manufacturing. *J Mate Proc Tech* 113, 301–5 (2001).

4. Kozac J, Rajurkar KP, Makkar Y. Selected Problems of Micro Electrochemical Machining. *Int J Mach Tool Manuf* 149, 426–31 (2004).

5. Hewidy MS, Ebeidb SJ, Rajurkarc KP, El-Saftia MF. Electrochemical Machining Under Orbital Motion Conditions. *J Mater Process Technol* 109, 339–46 (2001).

6. Bhattacharyya B, Munda J. Experimental Investigation into Electrochemical Micromachining (EMM) Process. *J Mater Process Technol* 140, 287–91 (2003).

7. Kim BH, Ryu SH, Choi DK, Chu CN. Micro Electrochemical Milling. *J Micromechan Microeng* 15(1), 124–9 (2005).

8. De Silvam AKM, McGeough JA. Process Monitoring of Electrochemical Micromachining. *J Mate Proc Tech* 76, 165–9 (1998).

9. Masuzawa T, Tsukamoto J, Fujino M. Drilling of Deep Microholes by EDM. *CIRP Ann* 38, 195–8 (1998).

10. Yu ZY, Masuzawa T, Fujino M. Micro-EDM for Three-Dimensional Cavities Development of Uniform Wear Method. *CIRP Ann* 47, 169–72 (1997).

11. Jain NK, Jain VK. Optimization of Electro-Chemical Machining Process Parameters Using Genetic Algorithms. *Mach Sci Technol* 11(2), 235–58 (2007).

12. Mukherjee SK, Srivastava PK, Kumar S. Effect of Over Voltage on Material Removal Rate During Electrochemical Machining. *J Sci Eng* 8(1), 1214–28 (2005).

13. Pawar A, Kamble D, Ghorpade RR. Overview on Electro-Chemical Machining of Super Alloys. *Mater Today: Proceed* 46, 696–700 (2021).

14. Lutey AH, Jing H, Romoli L, Kunieda M. Electrolyte Jet Machining (EJM) of Antibacterial Surfaces. *Prec Eng* 70, 145–54 (2021).

15. Bal KS, Nair AM, Dey D, Singh AK, Choudhury AR. Multi-objective Optimisation of Electro Jet Drilling Process Parameters for Machining of Crater in High-Speed Steel Using Grey Relational Analysis. In *Advances in Unconventional Machining and Composites* (pp. 385–95). Springer, Singapore (2020).

16. Goel H, Pandey PM. Performance Evaluation of Different Variants of Jet Electrochemical Micro-Drilling Process. *Proceed Inst Mech Eng, Part B: J Eng Manuf* 232(3), 451–64 (2018).

17. Goel H, Pandey PM. Experimental Investigations into the Ultrasonic Assisted Jet Electrochemical Micro-Drilling Process. *Mater Manuf Process*, 32(13), 1547–56 (2017).

18. Goel H, Pandey PM. Experimental Investigations and Statistical Modeling of Ultrasonic Assisted Jet Electrochemical Micro-Drilling Process with Pulsed DC. *J Adv Manuf Sys* 18(03), 413–34 (2019).

19. Sen M, Shan HS. A Review of Electrochemical Macro to Micro-Hole Drilling Processes. *Int J Mach Tool Manuf* 45, 137–52 (2005).

20. Ryu SH. Micro Fabrication by Electrochemical Process in Citric Acid Electrolyte. *Manuf Eng* 209, 2831–7 (2009).

21. Zhang Z, Zhu D. Experimental Research on the Localized Electrochemical Micro-Machining. *Russ J Electrochem* 44(8), 926–30 (2008).

22. Schaller T, Bohn L, Mayer J, Schubert K. Microstructure Grooves with a Width of Less Than 50 µm Cut With Ground Hard Metal Micro End Mills. *Prec Eng* 23, 229–35 (1999).

23. Rajurkar KP, Levy G, Malshe A, Sundaram MM, McGeough J, Hu X, Resnick R, DeSilva A. Micro and Nano Machining by Electro-Physical and Chemical Processes. *Annal CIRP* 55, 643–66 (2006).

24. Bhattacharyya B, Malapati M, Munda J. Experimental Study on Electrochemical Micromachining. *J Mater Process Technol* 169, 485–92 (2005).

25. Kurita T, Chikamori K, Kubota S, Hattori M. A Study of Three-Dimensional Shape Machining with an ECmM System. *Int J Mach Tool Manuf* 46, 1311–18 (2006).

26. Chikamori K. Possibilities of Electrochemical Micromachining. *Int J Jap Soc Prec Eng* 32 (1), 37–38 (1998).

27. Sharma A, Kumar V, Babbar A, Dhawan V, Kotecha K, Prakash C. Experimental Investigation and Optimization of Electric Discharge Machining Process Parameters Using Grey-Fuzzy-Based Hybrid Techniques. *Materials* 14(19), 5820 (2021 Jan).

28. Prakash C, Kumar V, Mistri A, Uppal AS, Babbar A, Pathri BP, Mago J, Sharma A, Singh S, Wu LY, Zheng HY. Investigation of Functionally Graded Adherents on Failure of Socket Joint of FRP Composite Tubes. *Materials* 14(21), 6365 (2021 Jan).

29. Babbar A, Jain V, Gupta D, Prakash C. Experimental Investigation and Parametric Optimization of Neurosurgical Bone Grinding Under Bio-Mimic Environment. *Surf Rev Lett* 28, 2141005 (2021 Jul).

30. Babbar A, Jain V, Gupta D, Agrawal D, Prakash C, Singh S, Wu LY, Zheng HY, Królczyk G, Bogdan-Chudy M. Experimental Analysis of Wear and Multi-Shape Burr Loading During Neurosurgical Bone Grinding *J Mater Res Technol* 12, 15–28 (2021 Feb 24).

31. Babbar A, Jain V, Gupta D, Agrawal D. Histological Evaluation of Thermal Damage to Osteocytes: A Comparative Study of Conventional and Ultrasonic-Assisted Bone Grinding. *Med Eng Phys* 90, 1–8 (2021 Feb 16).

32. Babbar A, Jain V, Gupta D, Agrawal D. Finite Element Simulation and Integration of CEM43°C and Arrhenius Models for Ultrasonic-Assisted Skull Bone Grinding: A Thermal Dose Model. *Med Eng Phys* 90, 9–22 (2021 Feb 16).

33. Babbar A, Prakash C, Singh S, Gupta MK, Mia M, Pruncu CI. Application of Hybrid Nature-Inspired Algorithm: Single and Bi-Objective Constrained Optimization of Magnetic Abrasive Finishing Process Parameters. *J Mater Res Technol* 9(4), 7961–74 (2020 Jul 1).

34. Baraiya R, Babbar A, Jain V, Gupta D. In-situ Simultaneous Surface Finishing Using Abrasive Flow Machining via Novel Fixture. *J Manuf Process* 50, 266–78 (2020 Feb 1).

35. Singh S, Prakash C, Pramanik A, Basak A, Shabadi R, Królczyk G, Bogdan-Chudy M, Babbar A. Magneto-Rheological Fluid Assisted Abrasive Nanofinishing of β-Phase Ti-Nb-Ta-Zr Alloy: Parametric Appraisal and Corrosion Analysis. *Materials* 13(22), 5156 (2020 Jan).

36. Babbar A, Jain V, Gupta D. Preliminary Investigations of Rotary Ultrasonic Neurosurgical Bone Grinding Using Grey-Taguchi Optimization Methodology. *Grey Systems: Theory and Application* 10, 479–493 (2020 Jun 23).

37. Babbar A, Jain V, Gupta D. In Vivo Evaluation of Machining Forces, Torque, and Bone Quality During Skull Bone Grinding. *Proceed Inst Mech Eng, Part H: J Eng Manuf* 234, 626–638 (2020 Mar 17).

38. Babbar A, Jain V, Gupta D. Thermogenesis Mitigation Using Ultrasonic Actuation During Bone Grinding: A Hybrid Approach Using CEM43°C and Arrhenius Model. *J Braz Soc Mech Sci Eng* 41(10), 401 (2019 Oct 1). (SCI IF = 1.755)

39. Babbar A, Sharma A, Jain V, Jain AK. Rotary Ultrasonic Milling of C/SiC Composites Fabricated Using Chemical Vapor Infiltration and Needling Technique. *Mater Res Express* 6, 085607 (2019 Apr 24). ISSN: 2053-1591.

40. Singh G, Babbar A, Jain V, Gupta D. Comparative Statement for Diametric Delamination in Drilling of Cortical Bone with Conventional and Ultrasonic Assisted Drilling Techniques. *J Orthop* 25, 53–8 (2021 May 1).

41. Babbar A, Sharma A, Singh P. Multi-Objective Optimization of Magnetic Abrasive Finishing Using Grey Relational Analysis. *Mater Today: Proc* 50, 570–5, (2022 Jan 1).

42. Sharma A, Kalsia M, Uppal AS, Babbar A, Dhawan V. Machining of Hard and Brittle Materials: A Comprehensive Review. *Mater Today: Proc* 50, 1048–52 (2022 Jan 1).

43. Babbar A, Sharma A, Bansal S, Mago J, Toor V. Potential Applications of Three-Dimensional Printing for Anatomical Simulations and Surgical Planning. *Mater Today: Proc* 33, 1558–61 (2020 Jan 1).

44. Babbar A, Jain V, Gupta D. Thermo-Mechanical Aspects and Temperature Measurement Techniques of Bone Grinding. *Mater Today: Proc* 33, 1458–62 (2020 Jan 1).

45. Sharma A, Babbar A, Jain V, Gupta D. Enhancement of Surface Roughness for Brittle Material During Rotary Ultrasonic Machining. In MATEC Web of Conferences 2018 (Vol. 249, p. 01006). EDP Sciences.

46. Babbar A, Singh P, Farwaha HS. Regression Model and Optimization of Magnetic Abrasive Finishing of Flat Brass Plate. *Indian J Sci Technol* 10, 1–7 (2017 Aug).

47. Babbar A, Singh P, Farwaha HS. Parametric Study of Magnetic Abrasive Finishing of UNS c26000 Flat Brass Plate. *Int J Adv Mechatronics Robot* 9, 83–9 (2017).

48. Babbar A, Sharma A, Chugh M. Application of Flexible Sintered Magnetic Abrasive Brush for Finishing of Brass Plate. *Optim Eng Res* 1(1), 36–47.

49. Babbar A, Jain V, Gupta D, Prakash C, Agrawal D (2022). Potential Application of CEM43° C and Arrhenius Model in Neurosurgical Bone Grinding. In: Chander Prakash, Sunpreet Singh, Aminesh Basak, J. Paulo Davim (eds) *Numerical Modelling and Optimization in Advanced Manufacturing Processes* (pp. 145–58). Springer, Cham.

50. Babbar A, Jain V, Gupta D (2019). Neurosurgical Bone Grinding. In: Prakash C et al. (eds) *Biomanufacturing*. Springer, Cham (Scopus indexed).

51. Sharma A, Grover V, Babbar A, Rani R (2020 Oct 30). A Trending Nonconventional Hybrid Finishing/Machining Process. In: Rupinder Singh, J. Paulo Davim (eds) *Non-Conventional Hybrid Machining Processes* (pp. 79–93). Routledge.

52. Babbar A, Jain V, Gupta D, Sharma A (2020 Oct 29). Fabrication of Microchannels Using Conventional and Hybrid Machining Processes. In: Atul Babbar, Vivek Jain, Dheeraj Gupta, Ankit Sharma (eds) *Non-Conventional Hybrid Machining Processes: Theory and Practice* (p. 37). CRC Press.

53. Babbar A, Jain V, Gupta D, Prakash C, Sharma A (2020 Oct 25). Fabrication and Machining Methods of Composites for Aerospace Applications. In: Chander Prakash, Sunpreet Singh, J. Paulo Davim (eds) *Characterization, Testing, Measurement, and Metrology* (pp. 109–24). CRC Press.

54. Sharma A, Jain V, Gupta D, Babbar A (2020 Oct 26). A Review Study on Miniaturization: A Boon or Curse. In: Chander Prakash, Sunpreet Singh, J. Paulo Davim (eds) *Advanced Manufacturing and Processing Technology* (pp. 111–31). CRC Press, Boca Raton.

3 Advanced Fabrication Techniques of Composites

A State of Art Review and Future Applications

Ankit Tiwari, K. Ponappa, and Puneet Tandon
Indian Institute of Information Technology, Design,
and Manufacturing, Jabalpur, India

3.1 INTRODUCTION

Composite materials have emerged as a revolutionary class of materials, with exceptional properties such as high strength, lightweight nature, and tailorable characteristics. These distinguishing characteristics have made composites highly desirable in industries ranging from aerospace and automotive to construction and sports equipment [1]. To fully realize the potential of composite materials, advanced fabrication techniques have been developed to improve their performance, manufacturing efficiency which have been studied by various researchers in the past [2–29], and the ability to create complex structures.

Traditional composite fabrication methods, such as hand lay-up and resin infusion, have been widely used and paved the way for composites' widespread adoption. However, these techniques frequently necessitate significant manual labour, have limitations in producing intricate shapes, and may result in variability in final product quality.

To address these challenges and push the boundaries of composite manufacturing, advanced fabrication techniques have been introduced. Automation, robotics, computer-aided design (CAD), and innovative processes are used in these techniques to achieve higher precision, better control over material placement, and shorter production times. They also enable the incorporation of functional features, customization, and the incorporation of advanced materials such as nanoparticles or nanofibres to improve the properties of composites.

The purpose of this chapter is to provide an in-depth examination of the advanced fabrication techniques used in composites manufacturing. It will go over the key benefits and drawbacks of each technique, as well as showcase their potential applications in various industries. Manufacturers can unlock new possibilities for

DOI: 10.1201/9781003427735-3

composite materials by understanding and utilizing these advanced techniques, driving innovation and pushing the boundaries of what is possible in terms of strength, weight reduction, and multifunctionality [30].

3.2 ADVANCED FABRICATION TECHNIQUES

A specialized method or process used in the manufacture of components or structures, typically involving the use of innovative approaches, equipment, and materials, is referred to as an advanced fabrication technique. These techniques go beyond traditional methods in order to improve the fabrication process's efficiency, precision, and performance [31].

Advanced fabrication techniques are critical for realizing the full potential of composite materials, enabling widespread adoption and expanding their applications across multiple industries. These techniques are designed to improve the manufacturing process, improve material properties, and achieve complex geometries with precision and efficiency [32].

3.3 TYPES OF ADVANCED FABRICATION TECHNIQUES

Here is a list of various advanced fabrication techniques utilized in different industries:

3.1. Automated Fibre Placement (AFP)
3.2. Tape Laying
3.3. Resin Transfer Moulding (RTM)
3.4. Additive Manufacturing (3D Printing)
3.5. Filament Winding
3.6. Vacuum-Assisted Resin Infusion (VARI)
3.7. Prepreg Fabrication
3.8. Automated Tape Bonding (ATB)
3.9. Resin Film Infusion (RFI)
3.10. Sheet Moulding Compound (SMC) Compression Moulding
3.11. Automated Deposition
3.12. Automated Tape Placement (ATP)
3.13. Continuous Tow Shearing
3.14. Electrospinning
3.15. Digital Fabrication
3.16. Laser Processing (Cutting, Welding, Cladding)
3.17. Injection Over Moulding
3.18. Hybrid Fabrication (combination of different techniques)
3.19. Powder Metallurgy
3.20. Hybrid Additive and Subtractive Manufacturing (SM)

Aerospace, automotive, marine, renewable energy, healthcare, electronics, and other industries use these advanced fabrication techniques. Each technique has distinct advantages and is tailored to specific needs, allowing components with improved properties, complex geometries, and increased manufacturing efficiency.

3.3.1 AUTOMATED FIBRE PLACEMENT

Automated Fibre Placement (AFP) is an advanced fabrication technique used in the manufacturing of composite structures. It involves the precise placement of continuous fibre reinforcements onto a mould or substrate using robotic systems. AFP enables the automated and highly controlled deposition of fibres, resulting in optimized fibre alignment and ply orientation [33].

The process begins with reimpregnated fibre tapes, also known as tows, which are fed into the AFP machine. The machine consists of a robotic arm or gantry system equipped with a fibre placement head. The head consists of multiple spools or creels that hold the tows and a series of rollers, guides, and heating elements to facilitate the placement process.

The AFP machine follows a predetermined path based on the design specifications, and the robotic arm moves the fibre placement head along this path. The tows are precisely positioned and consolidated onto the mould or substrate, layer by layer, with the help of tensioning devices and heat application [34]. The consolidation process ensures good adhesion between the fibre layers and proper resin flow (refers to Figure 3.1).

AFP finds applications in various industries, including aerospace, where it is used in the fabrication of aircraft wings, fuselages, and other structural components. It is also employed in the automotive industry for producing lightweight and high-strength body panels, chassis components, and interiors.

Overall, Automated Fibre Placement is a highly advanced and efficient technique that allows for precise fibre reinforcement placement, enabling the creation of composite structures with superior mechanical performance and complex geometries.

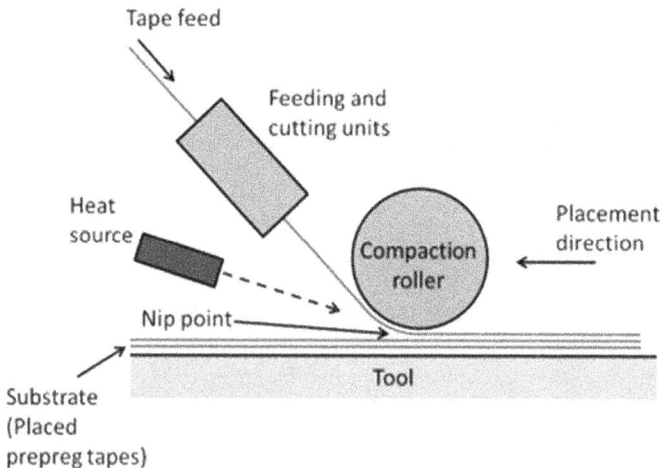

FIGURE 3.1 Automated fibre placement process [35].

a)   b)   Reinforcement feed

Tape collector   Tape feed

Cutter

Compaction
head   Resin bath
Mould   Heating elements

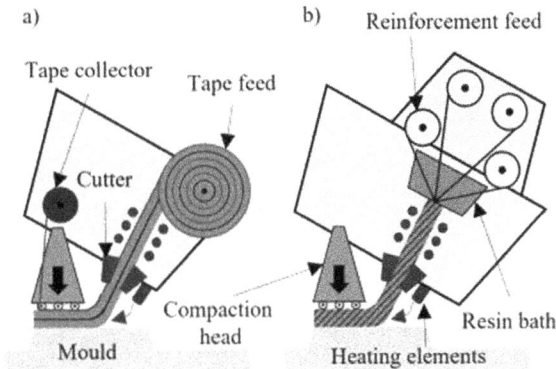

FIGURE 3.2   Automated tape lying [38].

3.3.2 TAPE LAYING

Tape laying is a complex fabrication technique used in the manufacture of composite materials. To create complicated composite structures, preimpregnated (prepreg)
tape is carefully positioned on a tool or mandrel. The tape is frequently composed of
resin-impregnated reinforcing fibre reinforcement, such as carbon or glass. The first
phase in the tape laying process is the preparation of the tool or mandrel, which is
generally coated with a release agent to aid in the removal of the cured composite
structure. Based on the design criteria, the prepreg tape is then cut into the desired
shapes and lengths [36].

The operator or automated system precisely aligns the fibres on the tool or mandrel during tape laying to obtain the correct fibre orientations and ply angles (refers
to Figure 3.2). Each layer of tape is applied one at a time, layer by layer, to form the
composite structure [37].

The tape is frequently placed strategically, with overlapping parts or specified
fibre orientations, to increase the structural integrity, strength, and performance of
the finished composite object. Pressure rollers or other consolidation methods may
be employed to ensure good contact between the tape and the underlying layers [39].

3.3.3 RESIN TRANSFER MOULDING

Resin Transfer Moulding (RTM) is a sophisticated fabrication technology used to
create composite materials. It is a closed-mould method in which liquid resin is
injected into a pre-formed dry fibre preform to make a composite product. The RTM
process begins with the creation of a two-part mould, often composed of metal or
composite materials, with a mould hole in the final part's intended shape. Inside the
mould cavity is a pre-formed dry fibre preforms which can be a stack of woven or
non-woven fibres [37].

The resin injection phase begins once the mould is closed. A low-viscosity liquid
resin, typically combined with additives such as catalysts or mould release agents, is
injected under pressure into the mould cavity. The resin travels through the fibre
preform, impregnating it and dissolving any trapped air or gases. The mould is built

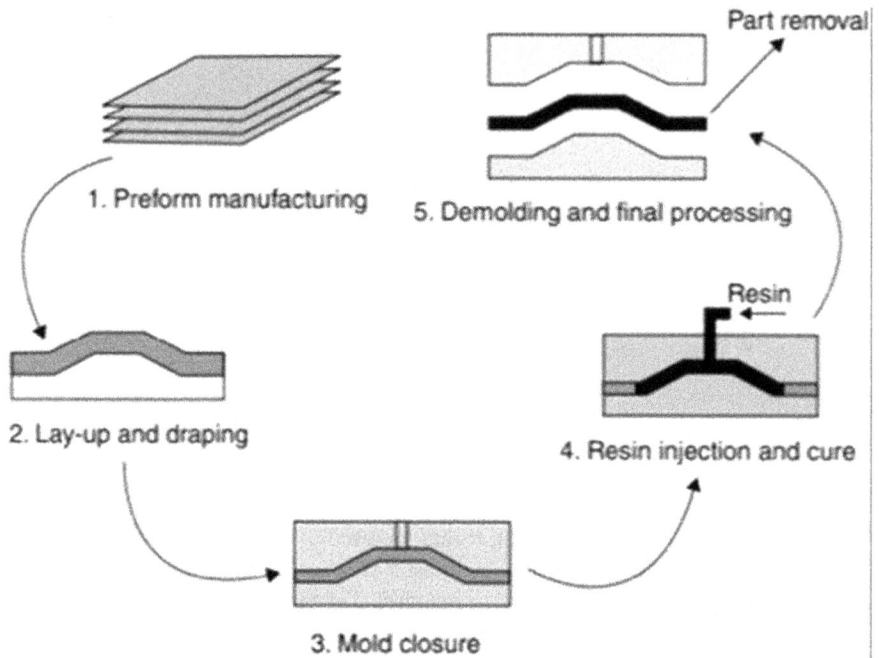

FIGURE 3.3   Resin transfer moulding process [40].

with channels or paths to direct the flow of resin and ensure uniform distribution (refers to Figure 3.3).

The mould is kept closed after the preform has been entirely saturated with resin to let the resin dry and set. Heat is frequently used in the curing process, either by heating the mould or by employing external heat sources. The resin goes through a chemical reaction that converts it from a liquid to a solid, joining the fibres together and making the composite part. The mould is opened and the hardened composite portion is removed once the resin has fully cured. To achieve the desired final shape and quality, further post-processing operations such as trimming, sanding, or surface treatments may be undertaken, depending on the unique requirements [41].

3.3.4   ADDITIVE MANUFACTURING (3D PRINTING)

Additive Manufacturing (AM), often known as 3D printing, is a sophisticated production technology that allows for the layer-by-layer creation of three-dimensional things. Unlike SM, which involves removing material from a solid block, AM constructs items by depositing material in a controlled manner. The AM process starts with the production of a digital 3D model with CAD software. The model is then split into thin cross-sectional layers that serve as 3D printer instructions [42].

The printer deposits material layer by layer during printing, following the instructions from the sliced model. Polymers, metals, ceramics, and composites are among the materials that can be utilized in AM. Depending on the 3D printing process, the material

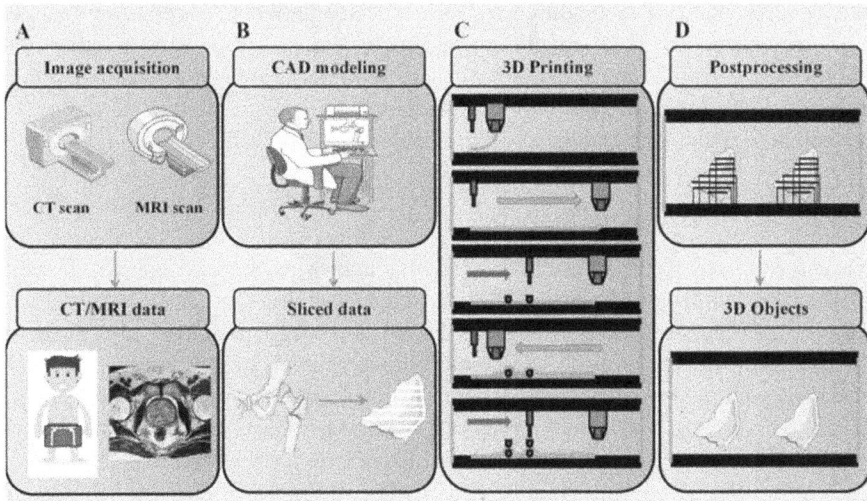

FIGURE 3.4 3D printing [43].

is typically in the form of filament, powder, or liquid. The printer carefully regulates the material deposition, sometimes using a nozzle or a laser, to solidify or fuse the material together, layer by layer, in accordance with the design specifications. This procedure is repeated until the complete item has been constructed (refers to Figure 3.4).

Design freedom, quick prototyping, and the capacity to construct complicated geometries that would be difficult or impossible to produce with traditional production methods are all advantages of AM. It also allows for customization and on-demand production, which lowers waste and inventory costs. However, as compared to traditional manufacturing techniques, AM may have limits in terms of material characteristics, surface polish, and production speed. The material and printing conditions used can have a considerable impact on the final quality and qualities of the printed object.

3.3.5 FILAMENT WINDING

Filament winding is a complex manufacturing technique used to make composite structures, specifically cylindrical or tubular components. It entails precisely positioning continuous reinforcing fibres, frequently in the form of a filament, onto a rotating mandrel or mould. The first step in the filament winding process is the creation of the mandrel or mould, which is typically made of metal or composite materials. The mandrel is connected to a rotating spindle, which provides rotation during the winding process [44].

The reinforcing filament, which is typically made of carbon fibre or fibreglass, is then passed through a resin bath or impregnation system to coat it with a liquid resin. The resin-coated filament is then directed onto the revolving mandrel in a predetermined pattern and tension. Depending on the desired structural properties of the component, the filament is precisely laid down in a helical or hoop direction while the mandrel rotates. The winding process is repeated until the desired layer count or thickness is reached. Throughout the winding process, pressure rollers or other consolidation mechanisms may be used to ensure good filament layer adhesion and consolidation. For consistent and high-quality winding, the resin impregnation process, as well as filament tension and speed management, are critical.

3.3.6 VACUUM-ASSISTED RESIN INFUSION

Vacuum-Assisted Resin Infusion (VARI), also known as Vacuum Infusion Process (VIP) or Resin Infusion Moulding (RIM), is a composite structure construction technology. Vacuum pressure is used to manage the infusion of resin into a dry fibre preform. VARI begins with the creation of a mould or tool, which is often constructed of hard materials such as fibreglass, metal, or composite materials. To assist the removal of the cured composite part, the mould is prepared with release agents or mould release films.

The mould is then covered with a dry fibre preform made of reinforcement fibres such as carbon, glass, or aramid. The preform is designed to conform to the final composite part's desired shape and structure. A vacuum bag or a flexible membrane is tightly wrapped around the mould and preform after it has been appropriately positioned. The vacuum bag is attached to a vacuum source to expel the air and create a vacuum pressure. The resin is put into the mould cavity after the vacuum is established. The resin is frequently a low-viscosity liquid that has been combined with additives such as catalysts or mould release agents. The vacuum pressure pushes the resin into the dry preform, impregnating the fibres and displacing any trapped air or gases (refers to Figure 3.5).

The resin infusion process is repeated until the entire preform has been saturated with resin. The vacuum pressure ensures even resin distribution throughout the preform and promotes thorough impregnation. Following the completion of the resin infusion, the part is allowed to cure and solidify. Curing is typically accomplished through heat, either through ambient curing or through the use of heat sources such as ovens or autoclaves. The curing time and temperature are determined by the resin system used. The vacuum bag is removed after the resin has fully cured, and the solidified composite part is demoulded. To achieve the desired shape and appearance, additional post-processing steps such as trimming, sanding, or surface finishing may be performed.

3.3.7 PREPREG FABRICATION

Prepreg fabrication is a cutting-edge manufacturing technique used to create composite materials. Prepreg is a type of composite material in which reinforcement fibres like carbon, glass, or aramid are pre-impregnated with a precise amount of resin. This resin is typically partially cured or in the B-stage. The preparation of the reinforcement fibres is the first step in the prepreg fabrication process. These fibres are frequently woven into fabric or arranged in the form of a unidirectional tape. The resin is then impregnated into the fibres via a variety of methods, including hot melt impregnation, solvent impregnation, and RFI.

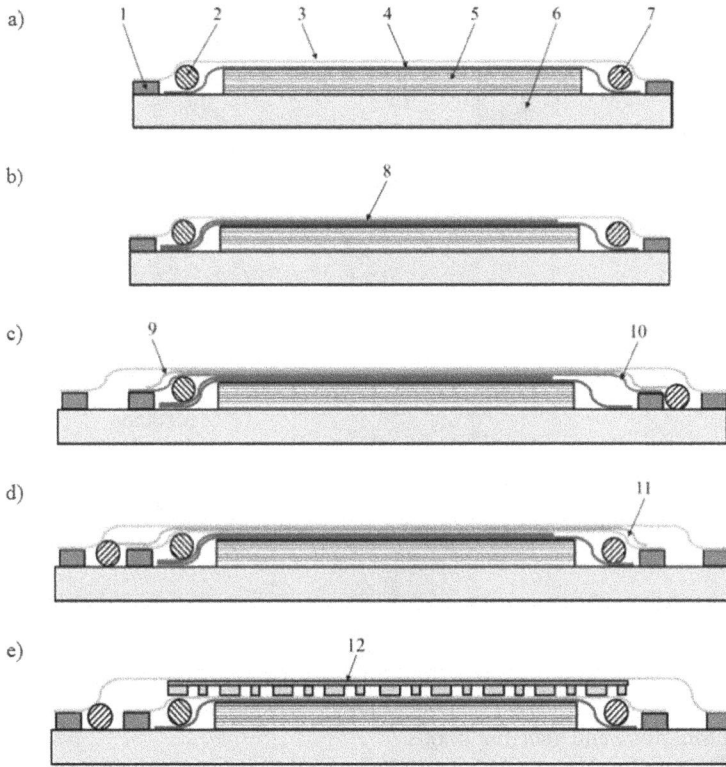

1: Sealant tape	2: Resin Inlet	3:Vacuum Bag	4: Release Fabric
5: Fibre Stack	6: Mould	7: Vacuum Outlet	8: Distribution media
9: Breather	10: Membrane	11: Inner Vacuum Bag	12: Pressure Distributor

FIGURE 3.5　Schematic of vacuum-assisted resin infusion [45].

Once impregnated, the fibres are typically stored in a freezer or refrigerated environment to extend their shelf life and prevent the resin from curing further. To create the desired composite structure, the prepreg material is cut into specific shapes or laid up in layers. The prepreg layers are frequently placed onto a tool or mould during the fabrication process. Depending on the desired shape of the final part, the tool can be a rigid surface or a mandrel. Layers are meticulously positioned to ensure proper fibre orientation and alignment (refers to Figure 3.6).

Following the application of the prepreg layers, the composite structure is subjected to heat and pressure to complete the curing process. This can be accomplished using techniques such as autoclave curing, oven curing, or press moulding. The heat activates the resin, allowing it to fully cure and produce a rigid composite part. Prepreg fabrication has a number of benefits, including precise resin control, consistent fibre volume fraction, and excellent mechanical properties. The fibres are impregnated with resin in a controlled environment, which ensures uniformity and eliminates the need for on-site resin mixing. This technique is widely used in

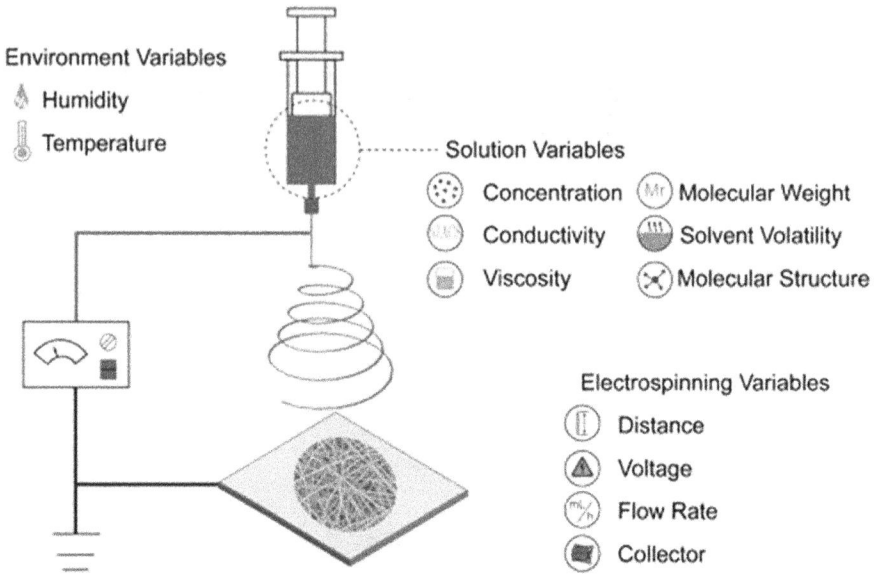

FIGURE 3.6   Prepreg fabrication [46].

high-performance composite component manufacturing in industries such as aerospace, automotive, and sporting goods.

3.3.8 AUTOMATED TAPE BONDING

ATB is a sophisticated fabrication technique for joining and bonding composite materials. It entails the automated placement and bonding of adhesive tapes onto composite structures in order to form strong and long-lasting bonds. The ATB procedure begins with the preparation of the composite parts to be joined. These components are typically made of composite panels or components made of carbon fibre, fibreglass, or aramid. To ensure proper adhesion, the surfaces of the parts are prepared by cleaning, sanding, or applying surface treatments.

Following that, adhesive tapes are precisely cut and positioned onto the prepared surfaces of the parts, which are typically made of high-performance materials such as epoxy or acrylic-based adhesives. This is accomplished through the use of automated equipment such as robotic arms or computer-controlled tape placement systems, which ensure precise tape placement (refers to Figure 3.7).

After positioning the tapes, pressure is applied to ensure proper contact between the tapes and the surfaces. This can be accomplished using a variety of techniques, such as rollers, vacuum bags, or mechanical clamps. By promoting adhesive flow and eliminating air voids, pressure aids in the bonding process. Following the initial bonding, the composite assembly is typically cured to fully activate the adhesive and achieve a strong bond. Depending on the adhesive and part requirements, the curing process may involve heat, pressure, or a combination of the two.

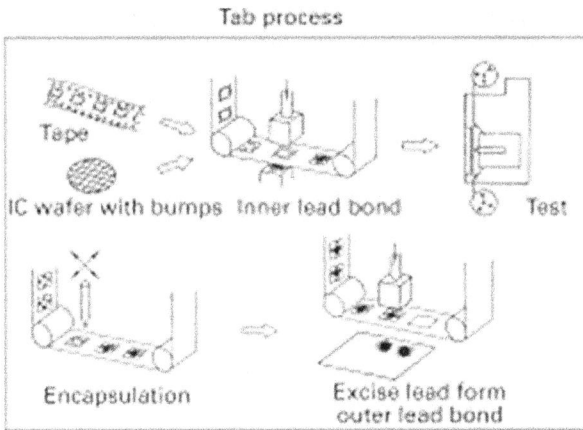

FIGURE 3.7 A single tube method for tape bonding [47].

The success of ATB, on the other hand, is dependent on proper surface preparation, adhesive selection, and precise control of process parameters. To achieve reliable and durable bonds, care must be taken to ensure proper alignment and contact between the adhesive tapes and the composite surfaces.

3.3.9 RESIN FILM INFUSION

RFI is a composite manufacturing process that involves infusing a resin film into a dry fibre reinforcement to form a composite part. Resin Transfer Moulding (RTM) or Resin Infusion are other terms for it. RFI typically starts with a dry fibre reinforcement, such as carbon fibre or fibreglass, that is preformed into the desired shape or placed in a mould. The dry fibre reinforcement is then covered with a resin film, which is a thin, flexible sheet of uncured resin.

To begin the infusion process, the mould containing the dry fibre reinforcement and resin film is typically sealed or enclosed, resulting in a closed system. This aids in the preservation of a controlled environment during the infusion process. The mould is frequently outfitted with resin inlet and outlet channels. The next step is to apply pressure and/or vacuum to encourage resin flow throughout the fibre reinforcement. The pressure difference aids in the penetration of the resin into the dry fibres, impregnating them and forming the composite structure. The resin wets out the fibres and fills the mould cavity as it flows through the reinforcement, conforming to the desired shape. During the infusion process, the resin film serves as a source of liquid resin, gradually releasing it into the reinforcement.

The part is typically cured under controlled temperature and pressure conditions after the mould cavity has been completely filled with resin and the reinforcement has been adequately impregnated. The curing process hardens the resin, resulting in a strong, rigid composite structure. RFI has a number of advantages in composite manufacturing. It enables the production of complex, lightweight parts with excellent fibre consolidation. The process also allows for the use of a variety of fibre reinforcements and resin systems, giving material selection flexibility. Furthermore, RFI can be heavily automated, resulting in increased process efficiency and consistency.

RFI is a composite manufacturing process that involves injecting a resin film into a dry fibre reinforcement to create high-quality composite parts. Because of its ability to produce strong, lightweight, and durable structures, it is widely used in a variety of industries, including aerospace, automotive, marine, and wind energy.

3.3.10 SHEET MOULDING COMPOUND COMPRESSION MOULDING

Sheet Moulding Compound (SMC) compression moulding is a method of producing high-strength composite parts. It entails pressing a pre-formed SMC sheet with a resin matrix and chopped fibre reinforcement into a heated mould cavity. The SMC material is a homogeneous mixture of thermosetting resins (such as polyester, vinyl ester, or epoxy) and chopped fibres (typically glass, carbon, or aramid). This mixture is then formed into thin, flat sheets, which are typically formed using a continuous sheeting process.

The SMC sheet is placed in a matched metal mould cavity, which is typically heated to a specific temperature, during the compression moulding process. The mould is created to ensure that the final part has the desired shape and features. The SMC sheet is positioned in the mould to ensure proper alignment and mould cavity coverage. Once the mould is closed, the SMC sheet is subjected to a combination of heat and pressure. Heat softens the resin matrix, allowing it to flow and conform to the shape of the mould, while pressure ensures that the material is evenly distributed and compacted within the mould cavity.

The combination of heat, pressure, and time causes the resin to go through a curing process known as polymerization in which it undergoes a chemical reaction and solidifies. As a result, the composite part is fully cured and has the desired shape, surface finish, and mechanical properties. Following the completion of the curing process, the mould is opened and the finished part is removed. Excess flash or trim is typically removed, and any additional operations, such as drilling or machining, may be performed to meet the final part specifications.

3.3.11 AUTOMATED DEPOSITION

The process of depositing materials layer by layer to build a three-dimensional object using automated systems is known as automated deposition, also known as automated material deposition or automated AM. It is a type of AM in which the object is built by adding material rather than removing it, as is the case with traditional SM processes.

A computer-controlled system, also known as a 3D printer or AM machine, precisely deposits material in a controlled manner according to a digital model or design in automated deposition. Extrusion, powder deposition, and liquid-based methods are some of the techniques that can be used for deposition. Extrusion-based deposition is one of the most common automated deposition techniques. A thermoplastic or thermosetting material is heated to a semi-liquid or molten state and extruded through a nozzle or a small opening in this method. The nozzle follows a predefined path, depositing material layer by layer to create the desired shape. Each layer that is deposited solidifies, allowing subsequent layers to be deposited on top.

Another technique used in automated deposition, particularly in metal AM processes, is powder deposition. A thin layer of metal powder is spread across a build platform in this method, and a focused energy source, such as a laser or electron beam, selectively melts and fuses the powder particles together in accordance with the digital design. After that, the build platform is lowered and a new layer of powder is applied on top. The process is repeated until the entire object has been constructed.

Liquid-based methods, such as stereolithography (SLA) or digital light processing (DLP), use light sources such as lasers or projectors to selectively cure a liquid resin. A vat contains the resin, and a platform is gradually lowered into the vat, allowing the cured layer to form. The process is repeated layer by layer until the object is finished.

Automation has found use in a variety of industries, including aerospace, automotive, healthcare, consumer goods, and architecture. It is used for prototyping, tooling, and end-use part production. With advancements in materials, process speed, and system capabilities, the technology continues to evolve, expanding its potential applications and driving manufacturing innovation.

3.3.12   AUTOMATED TAPE PLACEMENT

ATP is a composite structure manufacturing process that uses automated tape placement. It entails the automated placement of narrow strips or tapes of composite materials such as carbon fibre or fibreglass onto a mould or substrate to form layers and a composite part.

A computer-controlled machine, also known as an automated tape placement system, is used in ATP to precisely position and lay down the composite tapes onto the desired surface. A tape delivery system, a cutting mechanism, and a robotic arm or gantry system that moves the tape applicator comprise the machine. The procedure begins with the creation of a digital model or design of the desired part. This digital model is used to create a toolpath, which specifies the path and orientation of the tape as it is placed on the mould or substrate. The tape delivery system feeds the machine with composite tape, and the cutting mechanism cuts the tape to the desired length.

The tape applicator, which consists of a heated roller or compaction tool, is moved along the toolpath by the robotic arm or gantry system. As the tape passes through the applicator, it is heated, allowing it to conform to the shape of the mould or substrate. The heated roller applies pressure to the tape to ensure proper adhesion and consolidation with the underlying layers. ATP enables precise control over tape placement, allowing for the creation of complex geometries and contoured surfaces. To meet specific design requirements, the process can be tailored to optimize fibre orientation, ply thickness, and fibre volume fraction. Furthermore, ATP has a high deposition rate, allowing for the rapid fabrication of composite structures.

3.3.13   CONTINUOUS TOW SHEARING

Continuous tow shearing is a process used in composite material manufacturing to separate continuous fibre tows into narrower widths. It entails cutting or shearing wide fibre tows into smaller widths, allowing for greater material placement flexibility and improving the overall performance of the composite structure. A wide tow of continuous fibres, such as carbon fibre or fibreglass, is fed into a machine specifically designed for this purpose in continuous tow shearing. The machine includes a cutting mechanism, which can be a set of sharp blades or a shearing device, as well as a tension control system to ensure proper tow feeding. The cutting mechanism or shearing device applies a controlled force to the wide tow as it is fed into the machine, effectively dividing it into narrower strands or tapes. The desired specifications and requirements of the composite part being manufactured determine the width of the strands or tapes produced.

Continuous tow shearing is widely used in composite-material-intensive industries such as aerospace, automotive, sports and recreation, and wind energy. It is frequently used to improve the precision and efficiency of material placement in processes such as automated fibre placement (AFP), automated tape laying (ATL), and other automated manufacturing techniques. Overall, continuous tow shearing is important in the production of high-performance composite structures because it allows for better fibre alignment, conformability, and material optimization.

3.3.14 ELECTROSPINNING

Electrospinning is a fibre manufacturing technique that produces nanofibres with diameters ranging from a few nanometres to a few micrometres. It is a versatile and precise process that draws and elongates a polymer solution or melt into ultrafine fibres using electrostatic forces [48].

The electrospinning process usually consists of the following steps:

- Polymer Solution Preparation: To make a homogeneous solution, a polymer is dissolved in a suitable solvent. The polymer used is determined by the desired properties of the nanofibres, which include biocompatibility, mechanical strength, and electrical conductivity.
- Electrospinning Setup: The polymer solution is dispensed into a syringe or reservoir, which is attached to a metallic needle or a spinneret. The polymer solution is dispensed through a small opening in the spinneret.
- Electrostatic Field Generation: An electrostatic field is created by connecting a high voltage power supply to the spinneret. The applied voltage is determined by the polymer properties and the desired fibre diameter.
- Fibre Formation: The surface tension of the polymer solution causes a droplet to form at the tip of the spinneret as the voltage is applied. Surface tension is overcome by electrostatic forces, and a fine jet of polymer solution is ejected from the droplet.
- Fibre Drawing and Solidification: As it travels towards a grounded collector, electrostatic forces cause the jet of polymer solution to stretch and elongate. Solvent evaporation or cooling occurs during this flight path, solidifying the jet into a continuous nanofibre.
- Collection: The nanofibres are gathered on a grounded or rotating collector to form a nonwoven mat or an aligned nanofibre sheet (refers to Figure 3.8).

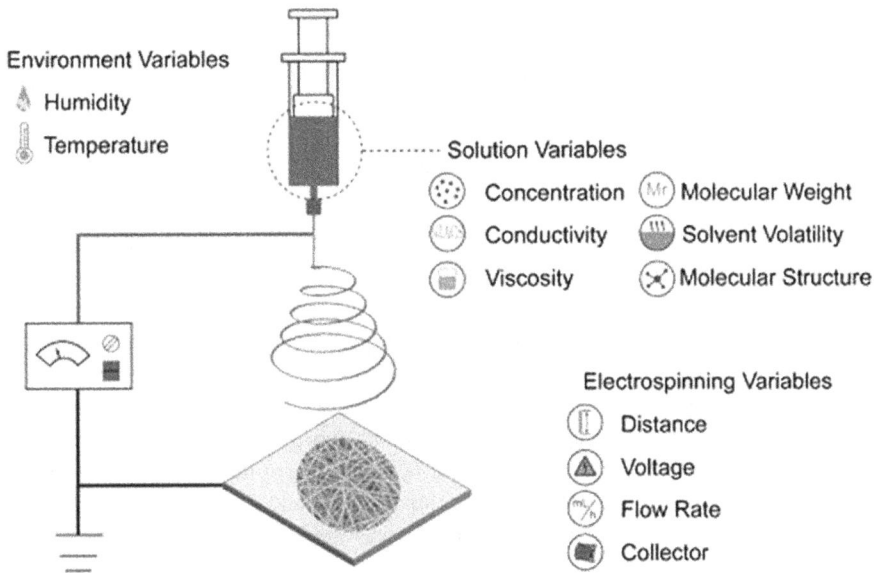

FIGURE 3.8 Electrospinning process [49].

3.3.15 DIGITAL FABRICATION

The use of computer-controlled machines and technologies to create physical objects from digital designs or models is referred to as digital fabrication. It refers to a variety of manufacturing processes that automate and streamline production by utilizing CAD and computer-aided manufacturing (CAM) techniques. The following key elements are involved in digital fabrication:

- Digital Design: The creation or acquisition of a digital design or model using CAD software is the first step in the process. The design specifies the specifications, dimensions, and geometry of the to-be-fabricated object.
- CAD/CAM Integration: The digital design is then linked to CAM software, which generates toolpaths and fabrication machine instructions. CAM software converts the digital design into machine-readable instructions for the fabrication process.
- Fabrication Process: Depending on the desired object and materials, various digital fabrication processes can be used. Among the most common processes are:
  a. 3D Printing/AM
  b. Computer Numerical Control (CNC) Machining
  c. Laser Cutting/Engraving
  d. Waterjet Cutting
- Fabrication Execution: The fabrication machine follows the toolpaths generated by the CAM software to produce the physical object. To achieve the desired result, the process may require multiple iterations or passes.

Several advantages exist between digital fabrication and traditional manufacturing methods. It allows for greater design flexibility, rapid prototyping, customization, and the creation of complex geometries. It also reduces material waste, enables design optimization through digital simulations, and makes the incorporation of digital sensors or electronics into objects easier.

3.3.16 LASER PROCESSING (CUTTING, WELDING, CLADDING)

Laser processing is the use of lasers in a variety of manufacturing processes such as cutting, welding, and cladding. Lasers produce a focused beam of high-intensity light that can be precisely controlled and directed to remove, join, or coat material.

- Laser Cutting: Laser cutting is a process where a laser beam is used to cut through materials by melting, burning, or vaporizing the material along a predetermined path. The laser beam is focused onto the workpiece, creating a high-energy density spot that rapidly heats and melts or vaporizes the material. A gas jet or air stream blows away the molten or vaporized material, leaving a clean and precise cut. Laser cutting is commonly used for materials such as metals, plastics, wood, and fabrics, and it finds applications in industries like automotive, aerospace, and signage.
- Laser Welding: Laser welding is a technique that joins or fuses materials together by using a laser beam. When the laser beam cools, it creates a localized heat source, melting the material and allowing it to solidify and form a strong bond. Laser welding allows for precise control of the energy input, resulting in weld seams that are narrow and controlled. It is widely used for joining metals, plastics, and even dissimilar materials in industries such as automotive, electronics, and medical devices.
- Laser Cladding: Laser cladding, also known as laser metal deposition, is a process that involves melting and fusing a powdered or wire-form material onto a substrate to form a coating or layer. The laser beam melts the material, which is either powdered or wired, and simultaneously melts it onto the substrate. Laser cladding is used to improve a material's surface properties, such as wear resistance and corrosion resistance, or to repair damaged parts. It is used in industries such as aerospace, oil and gas, and tooling.

In comparison to traditional manufacturing methods, laser processing has several advantages. It has high precision and accuracy, as well as the ability to process complex shapes or patterns. Lasers' non-contact nature reduces the possibility of material distortion or damage. Laser processing is also known for its fast-processing speeds, which can lead to higher productivity and efficiency.

It should be noted, however, that laser processing parameters such as power, focal point, and beam characteristics must be carefully controlled based on the material being processed and the desired outcome. Furthermore, due to the high-intensity light and potential generation of fumes or emissions during the process, laser processing may necessitate safety precautions. Overall, laser processing is a versatile and widely used technology that enables precise cutting, welding, and cladding operations with high efficiency and quality in a variety of industries.

3.3.17   INJECTION OVER MOULDING

Injection over moulding is a manufacturing process that combines two or more materials to form a single, integrated component. It is commonly used in the manufacture of complex parts that require a combination of different properties or functionalities. The following steps are typically involved in the injection over moulding process:

- Mould Preparation: A mould is created with cavities that will shape the final part. The mould is made up of two or more sections that can be opened and closed to allow for injection.
- First Material Injection: Using an injection moulding machine, the first material, also known as the substrate or base material, is injected into the mould cavities. The base material is usually a rigid plastic or metal that forms the part's core structure.
- Insert Placement: Following the injection of the first material, any necessary inserts, such as electronic components, metal parts, or other pre-formed elements, are placed into the mould cavity. These inserts will be incorporated into the final part.
- Second Material Injection: The second material, also known as over moulding material or overmould, is then injected over the first and inserts. The overmould material is typically a flexible or elastomeric polymer that adds grip, cushioning, or sealing properties.
- Cooling and Solidification: The mould is cooled after the injection of the overmould material to allow the materials to solidify and achieve the desired shape and properties.
- Mould Opening and Part Ejection: The mould is opened after the materials have solidified, and the finished part is ejected from the mould cavity (refer to Figure 3.9).

Injection moulding is used in a variety of industries, including automotive, electronics, medical devices, and consumer goods. Handles, grips, seals, connectors, and multi-material components are examples of over-moulded products. It should be noted that design and engineering considerations are critical in injection moulding. Material compatibility, part geometry, gate placement, and cooling must all be carefully considered in order to produce successful over-moulded parts.

3.3.18   HYBRID FABRICATION (COMBINATION OF DIFFERENT TECHNIQUES)

The use of multiple manufacturing techniques or processes to create a final product is referred to as hybrid fabrication. It entails combining multiple methods, such as AM, SM, or forming processes, to maximize the benefits of each and achieve a desired

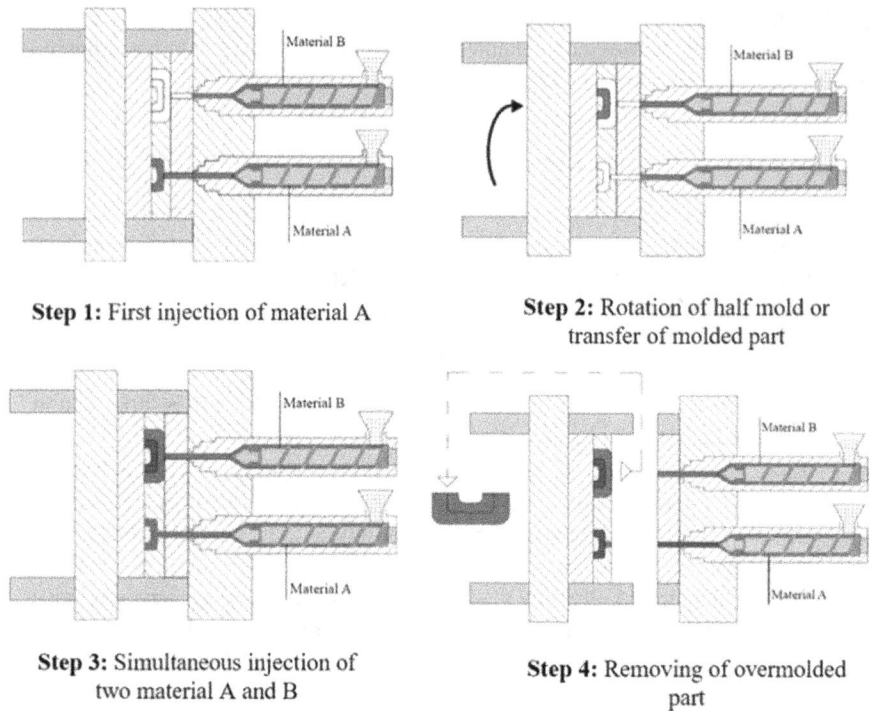

Step 1: First injection of material A

Step 2: Rotation of half mold or transfer of molded part

Step 3: Simultaneous injection of two material A and B

Step 4: Removing of overmolded part

FIGURE 3.9   Injection over moulding process [50].

result. By combining the strengths of various techniques, hybrid fabrication enables manufacturers to overcome limitations and optimize the production process.

Implementing hybrid fabrication, on the other hand, necessitates careful planning and consideration of the compatibility of various techniques, material properties, process parameters, and equipment integration. Furthermore, quality control and validation processes may need to be modified to ensure the final product's integrity. Overall, hybrid fabrication is a versatile approach to manufacturing that leverages the strengths of various techniques to achieve better results and meet specific manufacturing challenges. It is a developing field that will continue to evolve as manufacturing technologies advance.

3.3.19   POWDER METALLURGY

Powder metallurgy is a manufacturing process that involves the production of metal parts or components from fine metal powders. It is a versatile technique used to create complex shapes, achieve precise dimensions, and utilize a wide range of materials. The powder metallurgy process typically consists of the following steps:

* Powder Production
* Powder Blending
* Compaction

- Sintering
- Additional Operations:
  - Material Utilization
  - Complex Geometries
  - Material Properties
  - Cost Savings

Powder metallurgy is used in a wide range of industries, including automotive, aerospace, medicine, electronics, and consumer goods. It's used to make a variety of products, including gears, bearings, filters, cutting tools, magnets, and electrical contacts. To achieve the desired properties and dimensions, it is important to note that the powder metallurgy process necessitates careful control of parameters such as powder characteristics, compaction pressure, sintering temperature, and time. Furthermore, choosing the right powders, binders, and lubricants is critical to ensuring proper powder flow, uniform blending, and successful consolidation. Overall, powder metallurgy is a versatile and efficient manufacturing process for producing high-quality metal parts with numerous applications and advantages.

3.3.20 HYBRID ADDITIVE AND SUBTRACTIVE MANUFACTURING

The integration of both additive and SM processes within a single manufacturing system is referred to as hybrid additive and SM, also known as hybrid manufacturing or hybrid AM/SM. This method combines the benefits of both techniques to improve efficiency, flexibility, and part quality. The hybrid manufacturing process usually consists of the following steps:

- Design and Preparation: The part design is created or imported into CAD software which defines the integration of AM and SM features. After that, the CAD model is prepared for the hybrid manufacturing system, taking into account support structures, machining allowances, and surface finish requirements.
- Additive Manufacturing: The additive manufacturing process is started, and material is deposited layer by layer based on the CAD model. Various AM technologies, such as powder bed fusion, directed energy deposition, and material extrusion, can be used. The additive process uses the desired material to build up the initial shape of the part [51–56].
- Subtractive Manufacturing: The part may be subjected to subtractive processes after the additive manufacturing stage. This can include CNC machining which removes excess material to achieve final dimensions, smooth surfaces, and precise features. Milling, turning, drilling, and finishing are all examples of machining operations.
- Post-Processing and Finishing: Following the completion of the subtractive operations, the part may require post-processing and finishing steps such as surface treatment, polishing, coating, or heat treatment. These steps are taken to improve the surface quality, mechanical properties, and overall aesthetics of the part (refer to Figure 3.10).

(1) Start	(2) Additive process	(3) Milling	(4) Additive process

(5) Milling	(6) Additive process	(7) Milling	(8) Finish

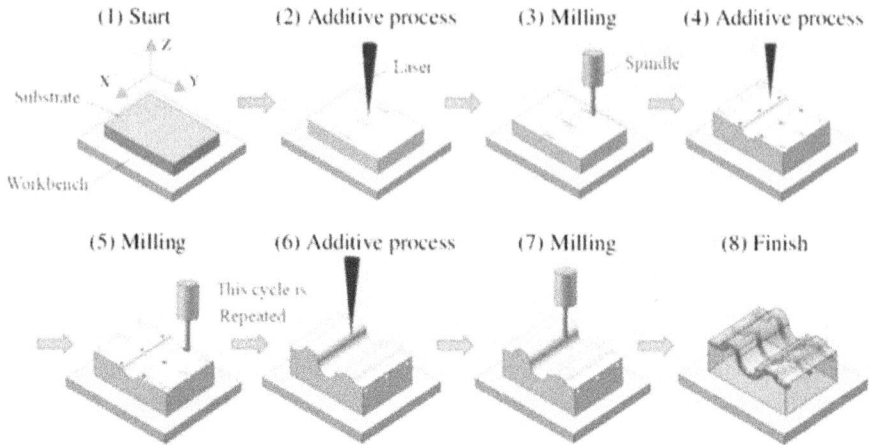

FIGURE 3.10 Schematic hybrid additive and subtractive manufacturing [57].

Hybrid additive and SM offers several benefits:

- Design Flexibility
- Improved Efficiency
- Material Flexibility
- Repair and Modification

Many industries use hybrid additive and SM, including aerospace, automotive, medical, and tooling. Prototypes, complex components, tooling inserts, customized parts, and even small production runs are all made with it. To ensure seamless coordination between the additive and subtractive processes, hybrid manufacturing systems must be implemented with careful consideration of equipment integration, process planning, toolpath generation, and control systems. Furthermore, effective workflow management and part inspection/validation techniques are required to ensure quality and accuracy throughout the manufacturing process [58].

3.4 COMPARISON BETWEEN VARIOUS ADVANCED FABRICATION TECHNIQUES

The comparative analysis of the various fabrication techniques along with the description, advantages, and disadvantages has been shown in Table 3.1.

3.5 CASE STUDIES

3.5.1 CASE STUDY 1: AUTOMATED FIBRE PLACEMENT IN AEROSPACE MANUFACTURING

Airbus' A350 XWB aircraft is a notable case study demonstrating the use of AFP in aerospace manufacturing. The A350 XWB (Extra Wide Body) is a long-range,

TABLE 3.1
Comparison between Various Advanced Fabrication Techniques

Technique	Description	Advantages	Disadvantages
Automated Fibre Placement (AFP)	Automated placement of fibre tapes for complex composite parts	High production rates, precise fibre placement	Expensive equipment, limited to flat or mildly curved parts
Tape Laying	Precise placement of preimpregnated tape for composite structures	Tailored fibre orientations, high fibre volume	Labour-intensive, limited to flat or mildly curved parts
Resin Transfer Moulding (RTM)	Injection of resin into a closed mould for composite parts	Good surface finish, complex geometries possible	Longer cycle times, cost of mould fabrication
Additive Manufacturing	Layer-by-layer 3D printing of composite parts	Design freedom, complex geometries possible	Limited strength compared to traditional methods
Filament Winding	Winding continuous fibre filaments onto a rotating mandrel	High strength, good fibre alignment	Limited to cylindrical or rotationally symmetric parts
Vacuum-Assisted Resin Infusion	Infusion of resin into a dry fibre preform under vacuum	Low-cost tooling, good resin distribution	Longer processing time, need for vacuum equipment
Prepreg Fabrication	Fabrication of composite sheets with preimpregnated fibres	Controlled resin content, consistent quality	High material cost, limited design flexibility
Automated Tape Bonding (ATB)	Automated bonding of composite tapes for structural components	High precision, reduced labour	Limited to specific applications and tape materials
Resin Film Infusion (RFI)	Infusion of resin film onto dry fibre preform under heat and pressure	Good fibre wet-out, consistent resin distribution	Limited to flat or mildly curved parts
SMC Compression Moulding	Compression moulding of fibre-reinforced sheets	High production rates, good surface finish	Limited design complexity, high tooling cost
Automated Deposition	Automated deposition of thermoplastic or thermoset material	High precision, reduced waste	Limited to specific materials and geometries
Automated Tape Placement (ATP)	Automated placement of composite tapes for structural components	High precision, reduced labour	Limited to specific applications and tape materials
Continuous Tow Shearing	Continuous shearing of fibre tows for composite reinforcement	High fibre alignment, improved properties	Limited to specific fibre types and applications
Electrospinning	Production of nanofibre mats by electrostatic deposition	High surface area, fine fibre diameter	Limited to thin mats or coatings
Digital Fabrication	Computer-controlled fabrication using various techniques	Customization, rapid prototyping	Equipment and material limitations

(Continued)

TABLE 3.1 (CONTINUED)
Comparison between Various Advanced Fabrication Techniques

Technique	Description	Advantages	Disadvantages
Laser Processing	Cutting, welding, or cladding using laser technology	High precision, minimal thermal damage	Limited to specific materials and geometries
Injection Overmoulding	Injection moulding of thermoplastic or thermoset over a substrate	Complex part integration, cost-effective	Limited to compatible materials and part designs
Hybrid Fabrication	Combination of different fabrication techniques	Tailored process for specific needs	Complex process planning and integration
Powder Metallurgy	Consolidation of powdered metals through compaction and sintering	Wide material selection, near-net shape forming	Limited to metallic materials, requires post-processing
Hybrid Additive and SM	Combination of 3D printing and machining	Design freedom, material flexibility	Increased process complexity and cost

twin-engine jetliner that uses advanced composite materials to reduce weight and fuel consumption.

The use of AFP technology was critical in the production of composite components for the A350 XWB. AFP machines were used by Airbus to precisely lay down carbon fibre tapes onto large mould, creating complex shapes and optimizing fibre orientations. The AFP process enabled precise fibre placement, lowering material waste and ensuring consistent quality. Airbus achieved significant weight savings in critical components such as the fuselage, wings, and tail sections by using AFP. The precise fibre placement resulted in optimized structural integrity, increasing the aircraft's overall strength and durability.

Furthermore, the use of composites allowed the A350 XWB to achieve greater fuel efficiency than traditional aluminium-intensive aircraft designs.

3.5.2 Case Study 2: Additive Manufacturing in Automotive Production

AM, also known as 3D printing, has found use in the automotive industry for rapid prototyping, tooling production, and even end-use part manufacturing. Ford Motor Company and the production of the Ford Shelby Mustang GT500 serve as an example.

Ford created lightweight composite intake manifolds for the GT500 using additive manufacturing technologies, specifically selective laser sintering (SLS). SLS 3D printing enabled the creation of intricate, optimized geometries that were previously impossible to achieve using traditional manufacturing methods. The resulting composite intake manifold improved airflow characteristics and contributed to the GT500's high-performance capabilities.

Ford significantly reduced production time and costs, as well as material waste, by using additive manufacturing. The design flexibility provided by AM also facilitated the creation of custom, lightweight parts tailored to the vehicle's specific requirements.

This case study demonstrates how advanced fabrication techniques such as AM can improve performance, reduce weight, and increase efficiency in automotive manufacturing.

These case studies demonstrate how advanced fabrication techniques can be used in real-world scenarios. Whether it's AFP in aerospace or AM in automotive, these advanced techniques have proven their ability to transform industries by enabling the production of high-performance components with improved properties and optimized designs.

3.6 CONCLUSION

To summarize, advanced fabrication techniques have transformed composite material manufacturing by enabling the development of high-performance components with improved properties and complex geometries. AFP, AM, and other techniques have demonstrated the ability to push the boundaries of what is possible in terms of material performance, manufacturing efficiency, and customization.

AFP allows manufacturers to achieve precise fibre placement, optimized ply orientations, and reduced material waste, resulting in composite structures with greater strength, stiffness, and durability. AM has opened up new possibilities for design freedom, rapid prototyping, and the manufacture of customized, lightweight components with intricate geometries that were previously difficult or impossible to manufacture using traditional methods.

Furthermore, advanced fabrication techniques have enabled industries to investigate the incorporation of advanced materials, such as nanocomposites, with enhanced mechanical, thermal, and electrical properties. These techniques have aided in the development of lightweight, strong, and multifunctional composite materials used in a wide range of industries, including aerospace, automotive, healthcare, and renewable energy.

As research and development progresses, we can expect further refinement and innovation in advanced fabrication techniques, leading to even more efficient, versatile, and reliable manufacturing processes. Continued research and application of these techniques will catalyse further advances in the field of composite materials and contribute to the development of next-generation products that meet the demanding requirements of modern applications.

REFERENCES

[1] T.P.D. Rajan and B.C. Pai, Developments in processing of functionally gradient metals and metal-ceramic composites: A review, *Acta Metallurgica Sinica (English Letters)* 27 (2014), pp. 825–838.

[2] M. Mehara, C. Goswami, S. Ranjan Kumar, G. Singh and M. Kumar Wagdre, Performance evaluation of advanced armor materials, *Materials Today: Proceedings* 47 (2021), pp. 6039–6042.

[3] A.S. Uppal, A. Sharma, A. Babbar, K. Singh and A.K. Singh, Minimum quality lubricant (MQL) for ultraprecision machining of titanium nitride-coated carbide inserts: Sustainable Manufacturing process, *International Journal on Interactive Design and Manufacturing (IJIDeM)* (2023), pp. 1–12. https://doi.org/10.1007/s12008-023-01299-4

[4] A. Babbar, V. Jain, D. Gupta, K. Goyal, C. Prakash, K. Saxena et al., Investigation of infrared thermography of cortical bone grinding in neurosurgery, *Advances in Science and Technology Research Journal* 17 (2023), pp. 116–123.

[5] Y. Tian, C. Tian, J. Han, A. Babbar and B. Liu, Characteristics of grinding force and Kevlar deformation of novel body-armor-like abrasive tool, *The International Journal of Advanced Manufacturing Technology* 122 (2022), pp. 2019–2030.

[6] Z. Gu, Y. Tian, J. Han, C. Wei, A. Babbar and B. Liu, Characteristics of high-shear and low-pressure grinding for Inconel718 alloy with a novel super elastic composite abrasive tool, *The International Journal of Advanced Manufacturing Technology* 123 (2022), pp. 345–355.

[7] A. Sharma, A. Babbar, Y. Tian, B.P. Pathri, M. Gupta and R. Singh, Machining of ceramic materials: A state-of-the-art review, *International Journal on Interactive Design and Manufacturing (IJIDeM)* 17 (2022), 2891–2911.

[8] A. Babbar, V. Jain, D. Gupta and D. Agrawal, Histological evaluation of thermal damage to Osteocytes: A comparative study of conventional and ultrasonic-assisted bone grinding, *Medical Engineering & Physics* 90 (2021), pp. 1–8.

[9] A. Babbar, V. Jain, D. Gupta, C. Prakash and D. Agrawal, Potential application of CEM43 °C and Arrhenius model in neurosurgical bone grinding, Springer, Cham, 2022, pp. 145–158.

[10] A. Babbar, V. Jain, D. Gupta, D. Agrawal, C. Prakash, S. Singh et al., Experimental analysis of wear and multi-shape burr loading during neurosurgical bone grinding, *Journal of Materials Research and Technology* 12 (2021), pp. 15–28.

[11] A. Babbar, V. Jain, D. Gupta and D. Agrawal, Finite element simulation and integration of CEM43 °C and Arrhenius Models for ultrasonic-assisted skull bone grinding: A thermal dose model, *Medical Engineering & Physics* 90 (2021), pp. 9–22.

[12] A. Babbar, V. Jain and D. Gupta, In vivo evaluation of machining forces, torque, and bone quality during skull bone grinding, *Proceedings of the Institution of Mechanical Engineers, Part H: Journal of Engineering in Medicine* 234 (2020), pp. 626–638.

[13] A. Babbar, V. Jain and D. Gupta, Thermogenesis mitigation using ultrasonic actuation during bone grinding: A hybrid approach using CEM43°C and Arrhenius model, *Journal of the Brazilian Society of Mechanical Sciences and Engineering* 41 (2019), pp. 401.

[14] A. Babbar, A. Sharma and P. Singh, Multi-objective optimization of magnetic abrasive finishing using grey relational analysis, *Materials Today: Proceedings* 50 (2022), pp. 570–575.

[15] A. Sharma, M. Kalsia, A.S. Uppal, A. Babbar and V. Dhawan, Machining of hard and brittle materials: A comprehensive review, *Materials Today: Proceedings* 50 (2022), pp. 1048–1052.

[16] A. Sharma, V. Kumar, A. Babbar, V. Dhawan, K. Kotecha and C. Prakash, Experimental investigation and optimization of electric discharge machining process parameters using Grey-fuzzy-based hybrid techniques, *Materials* 14 (2021), p. 5820.

[17] A. Babbar, V. Jain, D. Gupta and C. Prakash, Experimental investigation and parametric optimization of neurosurgical bone grinding under bio-mimic environment, *Surface Review and Letters* 30 (2023), pp. 2141005.

[18] G. Singh, A. Babbar, V. Jain and D. Gupta, Comparative statement for diametric delamination in drilling of cortical bone with conventional and ultrasonic assisted drilling techniques, *Journal of Orthopaedics* 25 (2021), pp. 53–58.

[19] S. Singh, C. Prakash, A. Pramanik, A. Basak, R. Shabadi, G. Królczyk et al., Magneto-rheological fluid assisted abrasive nanofinishing of β-Phase Ti-Nb-Ta-Zr alloy: Parametric appraisal and corrosion analysis, *Materials* 13 (2020), pp. 5156.

[20] A. Sharma, V. Jain, D. Gupta and A. Babbar, A review study on miniaturization, in Chander Prakash, Sunpreet Singh, J. Paulo Davim (Eds) *Advanced Manufacturing and Processing Technology*, First edition, CRC Press, Boca Raton, FL, 2021, pp. 111–131.

[21] A. Babbar, V. Jain, D. Gupta, C. Prakash and A. Sharma, Fabrication and machining methods of composites for aerospace applications, in Chander Prakash, Sunpreet Singh, J. Paulo Davim (eds) *Characterization, Testing, Measurement, and Metrology*, First edition. CRC Press, Boca Raton, 2020, pp. 109–124.

[22] A. Babbar, V. Jain, D. Gupta, C. Prakash, S. Singh and A. Sharma, Effect of process parameters on cutting forces and osteonecrosis for orthopedic bone drilling applications, in Chander Prakash, Sunpreet Singh, J. Paulo Davim (eds) *Characterization, Testing, Measurement, and Metrology*, First edition, CRC Press, Boca Raton, 2020, pp. 93–108.

[23] A. Babbar, V. Jain, D. Gupta and A. Sharma, Fabrication of microchannels using conventional and hybrid machining processes, in Chander Prakash, Sunpreet Singh, J. Paulo Davim (eds) *Non-Conventional Hybrid Machining Processes*, First edition, CRC Press, Boca Raton, 2020, pp. 37–51.

[24] A. Sharma, V. Grover, A. Babbar and R. Rani, A trending nonconventional hybrid finishing/machining process, in *Non-Conventional Hybrid Machining Processes*, First edition, CRC Press, Boca Raton, 2020, pp. 79–93.

[25] R. Baraiya, A. Babbar, V. Jain and D. Gupta, In-situ simultaneous surface finishing using abrasive flow machining via novel fixture, *Journal of Manufacturing Processes* 50 (2020), pp. 266–278.

[26] A. Babbar, C. Prakash, S. Singh, M.K. Gupta, M. Mia and C.I. Pruncu, Application of hybrid nature-inspired algorithm: Single and bi-objective constrained optimization of magnetic abrasive finishing process parameters, *Journal of Materials Research and Technology* 9 (2020), pp. 7961–7974.

[27] A. Babbar, A. Sharma, V. Jain and A.K. Jain, Rotary ultrasonic milling of C/SiC composites fabricated using chemical vapor infiltration and needling technique, *Materials Research Express* 6 (2019), pp. 085607.

[28] A. Babbar, V. Jain and D. Gupta, Neurosurgical bone grinding, in Chander Prakash, Sunpreet Singh, Seeram Ramakrishna, B. S. Pabla, Sanjeev Puri, M. S. Uddin (eds) *Biomanufacturing*, Springer International Publishing, Cham, 2019, pp. 137–155.

[29] A. Sharma, A. Babbar, V. Jain and D. Gupta, Enhancement of surface roughness for brittle material during rotary ultrasonic machining, *MATEC Web of Conferences* 249 (2018), p. 01006.

[30] W.S. Ebhota and T.-C. Jen, Casting and applications of functionally graded metal matrix composites, in T. Vijayaram (ed) *Advanced Casting Technologies*, InTech, 2018.

[31] S. El-Hadad, H. Sato, E. Miura-Fujiwara and Y. Watanabe, Fabrication of Al-Al$_3$Ti/Ti$_3$Al functionally graded materials under a centrifugal force, *Materials* 3 (2010), pp. 4639–4656.

[32] J. Lambros, A. Narayanaswamy, M.H. Santare and G. Anlas, Manufacture and testing of a functionally graded material, *Journal of Engineering Materials and Technology, Transactions of the ASME* 121 (1999), pp. 488–493.

[33] W.N.M. Jamil, M.A. Aripin, Z. Sajuri, S. Abdullah, M.Z. Omar, M.F. Abdullah et al., Mechanical properties and microstructures of steel panels for laminated composites in armoured vehicles, *International Journal of Automotive and Mechanical Engineering* 13 (2016), pp. 3741–3753.

[34] I.G. Crouch, Laminated materials and layered structures, in *The Science of Armour Materials*, Elsevier, 2017, pp. 167–201.

[35] A. Air, M. Shamsuddoha and B. Gangadhara Prusty, A review of Type V composite pressure vessels and automated fibre placement based manufacturing, *Composites Part B: Engineering* 253 (2023), p. 110573.

[36] N. Radhika, J. Sasikumar, J.L. Sylesh and R. Kishore, Dry reciprocating wear and frictional behaviour of B4C reinforced functionally graded and homogenous aluminium matrix composites, *Journal of Materials Research and Technology* 9 (2020), pp. 1578–1592.

[37] G. Udupa, S.S. Rao and K.V. Gangadharan, Functionally graded composite materials: An overview, *Procedia Materials Science* 5 (2014), pp. 1291–1299.

[38] V. Lunetto, M. Galati, L. Settineri and L. Iuliano, Sustainability in the manufacturing of composite materials: A literature review and directions for future research, *Journal of Manufacturing Processes* 85 (2023), pp. 858–874.

[39] Bassiouny Saleh, Jinghua Jiang, Reham Fathi, Tareq Al-Hababi, Qiong Xu, Lisha Wang, Dan Song and Aibin Ma, 30 years of functionally graded materials: An overview of manufacturing methods, applications and future challenges, *Composites Part B: Engineering* 201 (2020), p. 108376.

[40] N. Razali, M.R. Mansor, G. Omar, S.A.F.S. Kamarulzaman, M.H. Zin and N. Razali, Out-of-autoclave as a sustainable composites manufacturing process for aerospace applications, *Design for Sustainability: Green Materials and Processes* (2021), pp. 395–413. https://doi.org/10.1016/B978-0-12-819482-9.00011-3

[41] F. Klein, A. Litnovsky, X. Tan, J. Gonzalez-Julian, M. Rasinski, C. Linsmeier et al., Smart alloys as armor material for DEMO: Overview of properties and joining to structural materials, *Fusion Engineering and Design* 166 (2021), 112272.

[42] A. Levy, A. Miriyev, A. Elliott, S.S. Babu and N. Frage, *Additive manufacturing of complex-shaped graded TiC/steel composites, Materials & Design* 118 (2017), pp. 198–203.

[43] M. Meng, J. Wang, H. Huang, X. Liu, J. Zhang and Z. Li, 3D printing metal implants in orthopedic surgery: Methods, applications and future prospects, *Journal of Orthopaedic Translation* 42 (2023), pp. 94–112.

[44] G.T. Kridli, P.A. Friedman and J.M. Boileau, Manufacturing processes for light alloys, in *Materials, Design and Manufacturing for Lightweight Vehicles*, Elsevier, 2020, pp. 267–320.

[45] Z. Zhong, B. Zhang, Y. Jin, H. Zhang, Y. Wang, J. Ye et al., Design and anti-penetration performance of TiB/Ti system functionally graded material armor fabricated by SPS combined with tape casting, *Ceramics International* 46 (2020), pp. 28244–28249.

[46] M. Jabbar and A. Nasreen, Composite fabrication and joining, in Yasir Nawab, S.M. Sapuan and Khubab Shaker (eds) *Composite Solutions for Ballistics*, Woodhead Publishing, 2021, pp. 177–197.

[47] Kang S. K., Tape-automated bonding: Materials and technologies, *emst* (2001), pp. 9088–9093. https://doi.org/10.1016/B0-08-043152-6/01640-5

[48] R.S. Parihar, S.G. Setti and R.K. Sahu, Recent advances in the manufacturing processes of functionally graded materials: A review, *IEEE Journal of Selected Topics in Quantum Electronics* 25 (2018), pp. 309–336.

[49] A. Rogina, Electrospinning process: Versatile preparation method for biodegradable and natural polymers and biocomposite systems applied in tissue engineering and drug delivery, *Applied Surface Science* 296 (2014), pp. 221–230.

[50] N. Aliyeva, H.S. Sas and B. Saner Okan, Recent developments on the overmolding process for the fabrication of thermoset and thermoplastic composites by the integration of nano/micron-scale reinforcements, *Composites Part A: Applied Science and Manufacturing* 149 (2021), p. 106525.

[51] N. Ranjan, R. Tyagi, R. Kumar and A. Babbar, 3D printing applications of thermo-responsive functional materials: A review, *Advances in Materials and Processing Technologies* (2023), pp. 1–17. https://doi.org/10.1080/2374068X.2023.2205669

[52] A. Babbar, A. Sharma, V. Jain and D. Gupta, *Additive Manufacturing Processes in Biomedical Engineering*, CRC Press, Boca Raton, 2022.

[53] V. Kumar, C. Prakash, A. Babbar, S. Choudhary, A. Sharma and A.S. Uppal, Additive manufacturing in biomedical engineering, *Additive Manufacturing Processes in Biomedical Engineering*, CRC Press, Boca Raton, 2022, pp. 143–164.

[54] B.P. Pathri, Mohd S. Khan and A. Babbar, Relevance of bio-inks for 3D bioprinting, *Additive Manufacturing Processes in Biomedical Engineering*, CRC Press, Boca Raton, 2022, pp. 81–98.

[55] A. Babbar, V. Jain, D. Gupta, A. Sharma, C. Prakash, V. Kumar et al., Additive manufacturing for the development of biological implants, scaffolds, and prosthetics, *Additive Manufacturing Processes in Biomedical Engineering*, CRC Press, Boca Raton, 2022, pp. 27–46.

[56] A. Babbar, Y. Tian, V. Kumar and A. Sharma, *3D Bioprinting in Biomedical Applications*, in *Additive Manufacturing of Polymers for Tissue Engineering*, CRC Press, Boca Raton, 2022, pp. 1–16.

[57] W. Du, Q. Bai and B. Zhang, A novel method for additive/subtractive hybrid manufacturing of metallic parts, *Procedia Manufacturing* 5 (2016), pp. 1018–1030.

[58] B.I. Saleh and M.H. Ahmed, Development of functionally graded tubes based on pure Al/Al$_2$O$_3$ metal matrix composites manufactured by centrifugal casting for automotive applications, *Metals and Materials International* 26 (2020), pp. 1430–1440.

4 Electro-Chemical Machining of Titanium Alloys and Various Composites for Efficient Machining Process
A Critical Review

P. Sudhakar Rao and Vikas Sharma
National Institute of Technical Teachers' Training and
Research (NITTTR), Chandigarh, India

Harish Kumar
National Institute of Technology (NITD), Delhi, India

Sri Phani Sushma
University College of Engineering Narasaraopet, JNTU
Kakinada, India

4.1 INTRODUCTION

Electrochemical machining (ECM) has been in the thought of assembling humankind throughout the previous few decades because of its recognized points of interest of unimportant instrument wear, mechanical unstressed machining, and no heat affected zone [1–3]. ECM can fabricate progressed surface quality and tranquil surface items. It is the unconventional machining measures which be utilized hard to machine materials, for example, composite steel, Ti alloys, super alloys and treated steel, and so forth ECM is portraying as switched electroplating measure [4–6]. This is the reason for this cycle which is extremely famous, yet outside these businesses additionally in favour of some different purpose like electroplating of various equipment [7, 8]. ECM offers great advantages over other conventional methods. ECM process assumes a vital function in the assembling of an assortment of parts going from machining of enormous bits of complex shapes to micro level. In hybrid process, the mechanism of two processes is applied concurrently or consecutively. Although, the performance of combined process is better

DOI: 10.1201/9781003427735-4

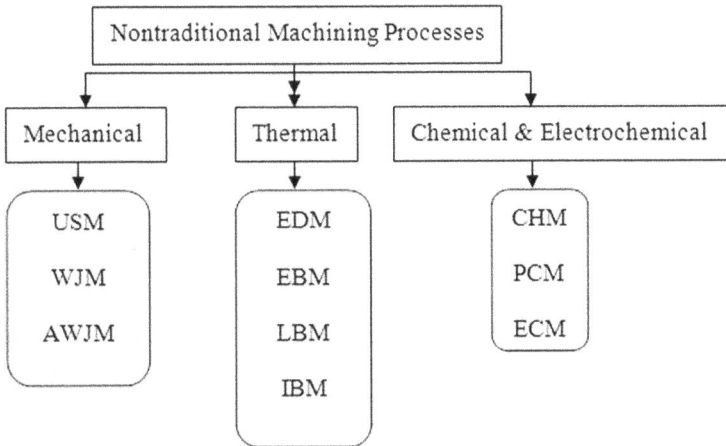

FIGURE 4.1   Classification of nontraditional machining processes.

as compared to the individual processes but hybridization increases the process complexity [9–11].

Complex shaped turbine blades are made of superalloys, and producing holes and slots in such alloy materials becomes problematic if the traditional processes are used. The higher production rate and economic requirements may demand the use of non-traditional (nonconventional) machining processes. Nonconventional (advanced) machining processes can be classified into three basic categories, i.e. mechanical, thermoelectric, and electrochemical and chemical machining processes. Figure 4.1 shows the classification of nontraditional machining processes.

Titanium is one of the most conductive materials which is very difficult to machine by traditional as well as non-traditional machining methods and has high potential among other superalloys having a high strength to weight ratio and low machinability. Titanium and its alloys are widely used in automotive industries, prosthetic devices, nuclear industries, chemical processing equipment, etc., due to their biological compatibility and ability to maintain strength at high temperatures [12, 13].

In this ECM is a very capable technology as it offers great advantages such as higher machining rate, machining wide range of materials, better precision, and control. The process is also cost effective and environment friendly.

ECM is restricted anodic disintegration measure in which high current is required between the tool and work piece throughout conductive liquid which is likewise called electrolyte [14–17]. It is non-contact process wherein empty space acquired is the imitation of hardware shape. Electrolyte is siphoned into the electrode gap between terminal hole (IEG) between apparatus and work piece. High current is flowing during this electrochemical circuit to break up metal particles from the work piece. The development of particles is joined by electrons stream, towards the path inverse to the positive current in the electrolyte. The mass disintegrated is straightforwardly relative to the amount of current stream into the circuit as per Faraday's law [18].

Nowadays high strength, low weight metallic and inter-metallic alloys are used for many applications. All ordinary techniques for machining like processing and turning are generally done so that instrument of hard material is utilized for scratching out the gentler material of the work piece [19, 20]. It is also uneconomical due to tool wear and low machining rates. As a consequence, some non-traditional machining techniques had to be investigated and developed during the last decades for producing better material removal rate (MRR). Non-conventional machining processes generally used are laser beam machining (LBM), electrical discharge machining (EDM) and ECM.

For the electrolysis to occur, an electric current is required to pass between two electrodes inserted into an electrolyte solution. Because of this, a chemical reaction occurs at the electrodes, which is recognized as an anodic or cathodic reaction or, in single word, electrochemical dissolution (ED) [21]. The basis for the EC machining is ED of the anodic work piece. Figure 4.2 shows the schematic of conventional ECM process. An electrolyte is being used to complete the electric circuit between the anodic workpiece and cathodic tool and prevents the anodic material from being deposited on the cathode. Hence, it can be called as a reverse process of electroplating.

In the initial two cycles, the material is taken out by dissolving due to high warm energy applied between the anodes [22, 23]. Despite the fact that, in these cycles, the material is disconnected without the contact of hardware and work piece, the heat affected zone (HAZ) and surface breaks are conceivable on the work piece.

Typically, a low voltage (7–25 V DC constant) is practical between the cathodes with a little hole (about 0.25 mm) keep up between them. Electrolyte (e.g., NaCl or NaNO$_3$ fluid arrangements) stream between the terminals with a high speed (30–60 m/s) to keep up the anodic disintegration just as to clear away the response items. Current streaming between the apparatus and the work piece through the terminals causes an exhausting activity to happen, which eliminates material from the work piece. Because of its capacity to machine hard-to-cut materials and confounded shapes without bending, scratches, burrs, and stress, ECM is as of now applied in the assembling of a wide scope of parts, for example, airplane turbine edges, careful inserts, bearing confines, merges and passes on, gunnery shots [24–27].

This process is likewise utilized for moulding and finishing activity in aviation and hardware businesses.

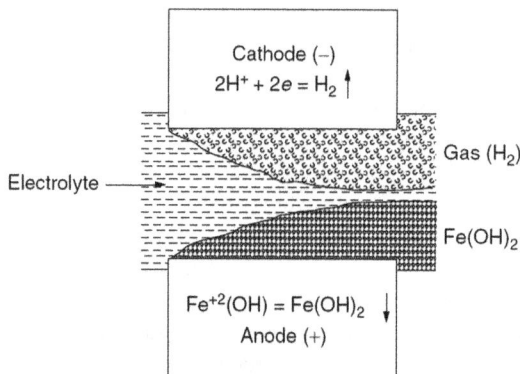

Cathode (–)

$2H^+ + 2e = H_2 \uparrow$

Gas (H$_2$)

Electrolyte

Fe(OH)$_2$

$Fe^{+2}(OH) = Fe(OH)_2 \downarrow$

Anode (+)

FIGURE 4.2   Electrochemical reactions during ECM process.

4.2   IMPORTANT POINTS ABOUT THE ECM PROCESS

Some of the important features of ECM process are listed below [28–30].

1. A potential contrast is kept up between the cathodes, thus particles existing in the electrolytes move towards the terminals.
2. Conventionally, emphatically charged particles are pulled in towards the cathode and negative particles are joined towards to the anode or stream of current starts in the electrolyte.
3. The wanted measure of metal is eliminated due to particle movement towards the apparatus.
4. For guarding the apparatus from harm, a constant flexibly of electrolyte is guaranteed by siphoning it at high weight ($15 kg/cm^2$).
5. In this cycle, the temperature created is low and no flash is delivered and along these lines there isn't any extent of metallurgical changes in the work.
6. In ECM, the device and work piece don't come in direct contact with one another, so immaterial wear is noticed.
7. Metal evacuation rate is high and voltage provided is low.
8. The metallic work piece isn't harmed because of warm burdens.

In ECM the dissolution products such as metal hydroxides, heat, and gas bubbles produced in the inter-electrode gap are removed by electrolysis. ECM process has the ability to produce complex cavities in high-strength materials and it helps in producing smoother surfaces [31–33]. Machining of different materials leads to the removal of material in the form of microchips [34, 35].

Tool wear in this process is nearly zero, since there is no direct contact between tool and workpiece. Based on the ECM principle, a number of industrial techniques have been developed. Figures 4.3 and 4.4 show the schematic diagram of ECM process and ECH process, respectively.

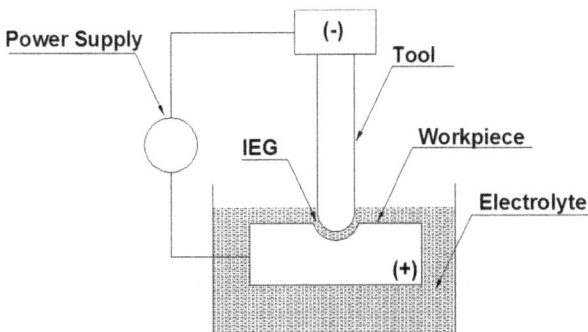

FIGURE 4.3   Schematic diagram of ECM process.

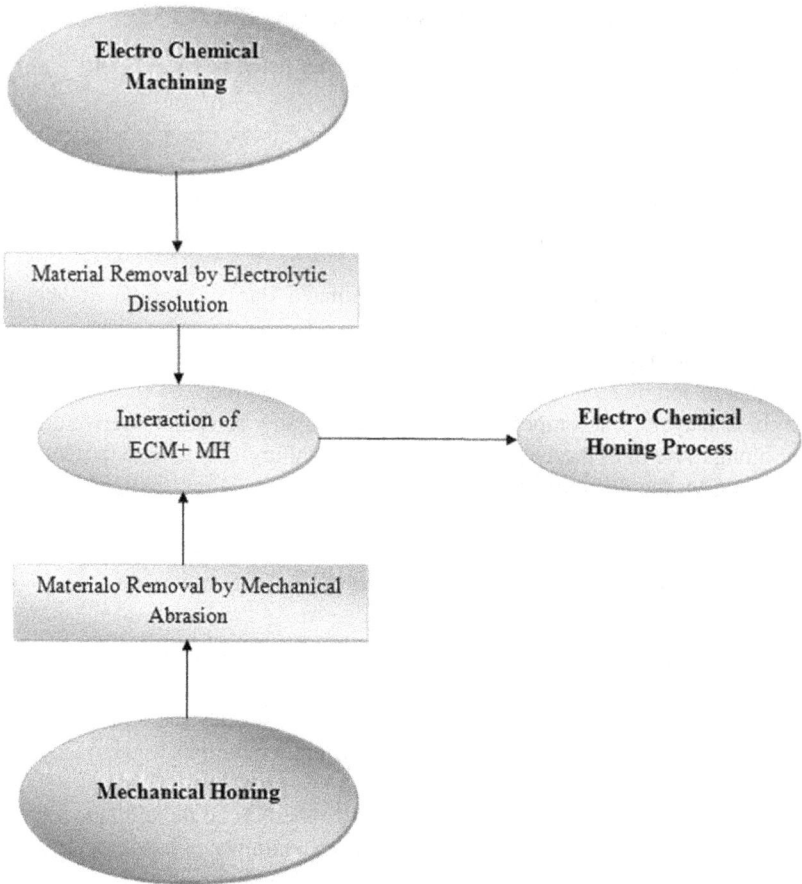

FIGURE 4.4 Fundamental concept of ECH: Combination of ECM and mechanical abrasion.

4.3 IMPORTANT PARAMETERS OF ECM PROCESS

4.3.1 IMPORTANT PROCESS PARAMETERS OF ECM PROCESS

The parameter selection process is based on different published papers of electrochemical honing. A few of those are given below. ECH is a hybrid process, therefore it includes the parameters of both ECM and conventional honing. The parameters can be divided into several categories like power related, electrolyte related, honing related and work piece related as described in Figure 4.5.

a. **Current**

The rate of material evacuation depends upon the current. The higher the current, the more the rate of material evacuation. This is usually seen to have some value because it unconstructively affects the completion and precision of the work item.

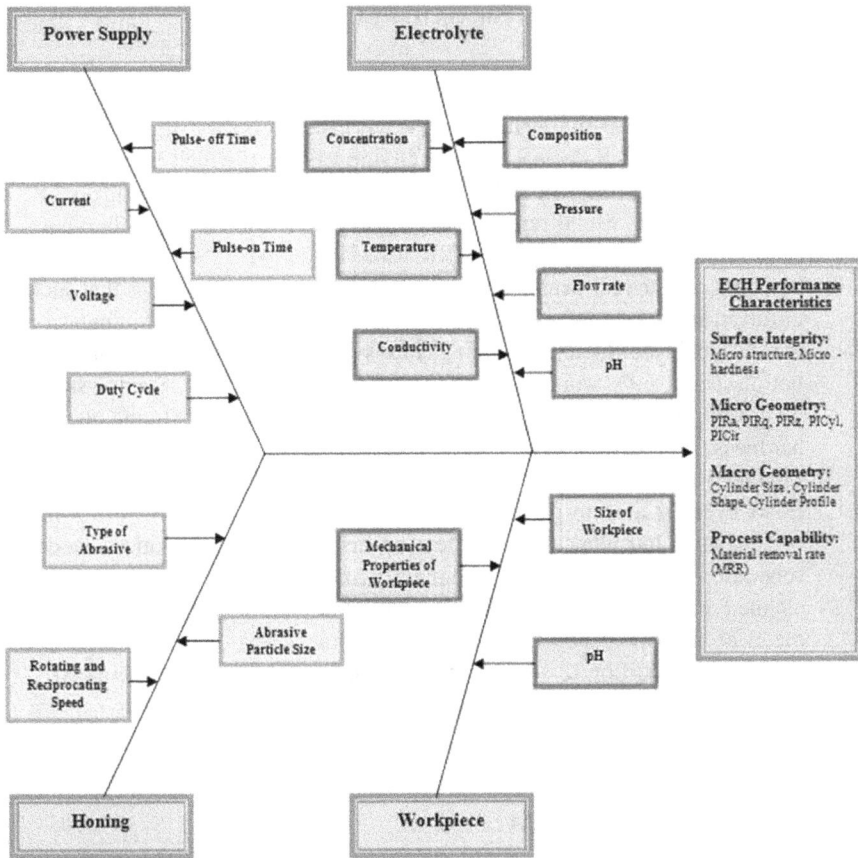

FIGURE 4.5   Process parameters.

b. **Tool Feed Rate**

In ECM measure, the hole is kept up to 0.01–0.07 mm among apparatus along with work piece. Intended for least hole, the electrical opposition among the device and work piece minute and most extreme flow and consequently greatest material is to be taken out. The instrument is taken care of towards the work piece contingent on how quick the material is to be eliminated. The instrument feed is to be constrained by water-powered chamber-based framework [36].

c. **Electrolyte and Its Concentration and Composition**

For the method of electrolysis to work, an electrolyte solution is necessary. In ECM, an electrolyte carries out three fundamental functions that are as follows:

i.   It carries high current to pass between tool and work piece.

ii. It dissipates maximum heat during in the machining process.

iii. It eliminates the results of machining response from cutting area.

4.3.2　IMPORTANT MACHINING PARAMETERS [37, 38]

a. **Material Removal Rate**

Material evacuation rate (MRR) is a component of feed rate that shows the measure of material eliminated from the work piece. As the distance between tool and workpiece decreases, current starts flowing which increases the material removal rate. In ECM the high feed rate produces great quality of surface finish. Metal removal rate depends on voltage, electrolyte concentration, and temperature as MRR value is low with low value of these parameters. The real aim of the metal evacuation rate technique is that such unfortunate surface effects that have occurred in standard machining steps are not achieved. Quiet machining, low system wear, and cancellation of hot damage to the workpiece are the principal advantages. Such systems do not impact mechanical properties such as yield quality, extreme quality, hardness, and ductility [39].

b. **Surface Finish**

ECM cycle is able to give surface completion 0.45 µm through turning of hardware and work piece. Defect on instrument face produces model scheduled the work piece. Consequently, the instrument surface must be cleaned before put into machining. The completion is better in harder material. For most beneficial surface harshness, cathode configuration, feed rate in addition to surface rectified added substances in the electrolyte be preferred. Low electrolytic temperature advances better surface roughness.

4.4　ADVANTAGES OF ECM

Some of the advantages of ECM process are as follows:

1. No tool wears during ECM process.
2. No burr formation on work piece.
3. High exterior finish can be achieved as a result of this process.
4. No thermal or mechanical stresses.
5. Hard conductive materials can be machined into complex shapes.
6. Fragile part can be machined easily as there is no stress involved [40].

4.5　APPLICATIONS OF ECM PROCESS

ECM finds its great application in different industries. Some of these include the following.

It contains the detailed discussions about process parameters, workpiece selection, parametric study to find out ranges and levels of input process parameters and fixed parameters, process performance characteristics and DOE technique to plan the experimental runs. The procedure and the measuring instruments utilized for quantifying the process performance parameters are also discussed in this chapter. The flow chart given in Figure 4.6 shows various steps involved during the optimization process.

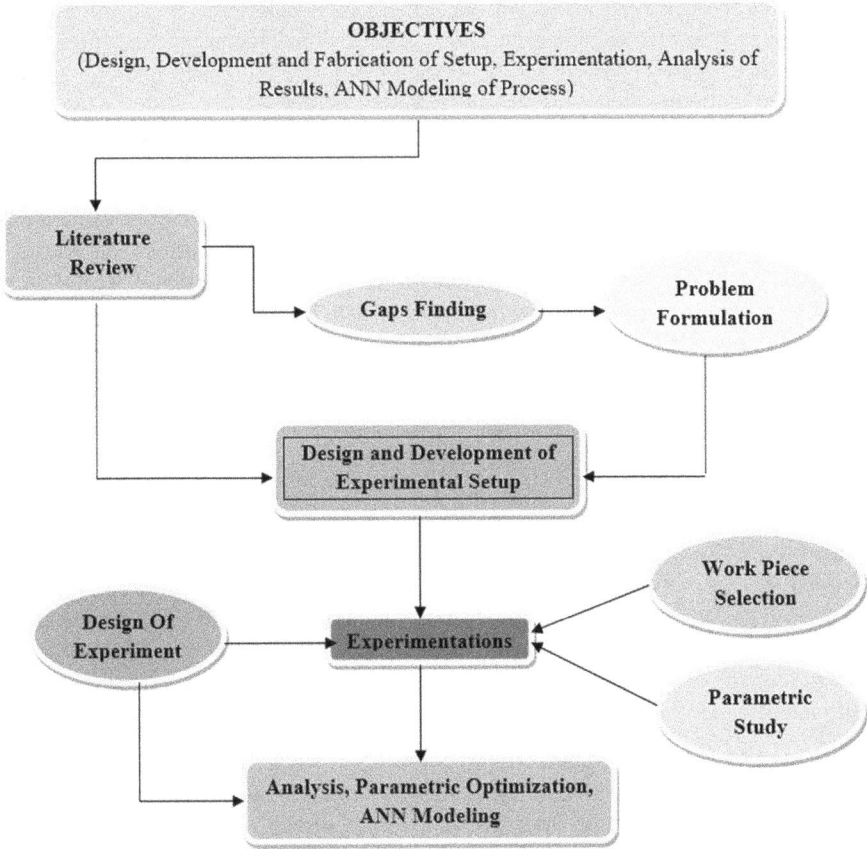

FIGURE 4.6   Flowchart to illustrate various steps of optimization process.

1. Automobile parts, aerospace, and jet engine parts.
2. Turbine blades and turbine nozzles in marine technology.
3. Quality demands are rising in the pharmaceutical industry.
4. Bio-medical components, surgical implants, etc.
5. Structural apparatus.
6. Cryogenic tankage.
7. Components for liquid fuel rocket engines.

4.6 STUDIES ON ELECTRO CHEMICAL MACHINING PROCESS OF TITANIUM ALLOY COMPOSITES

Baehre et al. [41] studied the electrochemical disintegration in watery potassium bromide, sodium chloride and sodium nitrate electrolytes with various pH values. The test results do that the disintegration experience is controlled by structure of the material and the electrolyte conditions. Furthermore, halides containing electrolytes lead to a higher material expulsion rate. On account of utilization $NaNO_3$ electrolyte, raised pH esteems progress the electrochemical disintegration of titanium composites. Paper norms of Ediss and Ecorr increment with increment titanium content while current thickness diminishes with the other one.

Chen et al. [42] Electrochemical micromachining of miniature dimple exhibits on the outside of Ti-6Al-4V with $NaNO_3$ electrolyte. Through-veil electrochemical micromachining (TMEMM) is a likely way to produce miniature dimple exhibits on metal surfaces. To secure excellent miniature dimple clusters, a beat current was working in TMEMM. The results show that machining boundaries of applied voltage of 24 V, beat obligation pattern of 10% and recurrence of 100 Hz were appropriate to improve the machining quality. With the $NaNO_3$ electrolyte, direct current was not suitable for producing miniature dimple clusters on the Ti-6Al-4V surface, for the explanation that it prompted genuine wanderer consumption in the veil covered locale.

Liu et al. [43] This paper manages test examination on ECM of gamma TiAL entombs metallic. Execution of gamma TiAL buries metallic form itself to be a sensible option for nickel composites in air motor applications. Symmetrical experiments were carried out to contemplate the enhancements of advancement boundaries, for example, applied voltage, terminal feed rate, electrolyte weight, and temperature on MRR, surface unpleasantness and machining hole in sodium chloride fluid arrangement. The result information of examinations was analysed by dim social investigation technique. The outcomes assigned that anode feed rate is the basic impact on MRR, SR, and MG, and the best factor mix was resolved.

Manikanadan et al. [44] The essential point of this work is to build up the working boundaries of electrochemical penetrating of titanium amalgam Ti-6Al-4V. cCntrol of different boundaries like feed rate, electrolyte stream rate, and convergence of electrolyte over material evacuation rate and overcut is examined. Taguchi's symmetrical exhibits, sign to commotion proportion, the examination of difference and dark social investigation are utilized to dissect the impact of these boundaries and locate the ideal cycle boundary level for multi-target reaction.

Ningsong et al. [45] wire electrochemical machining (WECM) is commonly utilized for work piece cutting under the circumstance of various thickness plates. In this paper, the pivotal electrolyte flushing is open to WECM for eliminating electrolysis items and restoring electrolyte. The machining process was carried out with Taguchi experimental design method.

Weinmann et al. [46] The disintegration conduct of two distinctive titanium amalgams in fluid electrolytes has been researched. After arrangement by bend dissolving and a warmth treatment, a broad portrayal of the amalgam by X-beam diffraction and energy dispersive X-beam spectroscopy has been performed. The electrochemical conduct of the composites in various electrolytes was examined utilizing various methods.

Kumar et al. [47] Ti-6Al-4V ELI (Grade 23) was machined with EDM machining measure taking heartbeat on schedule, voltage and current as the cycle boundaries. Surface completion and material evacuation pace of each occupation was determined. Surface morphology was done on the machined surface to check the conduct of the work piece. Multi-reaction Gray Relation Analysis (GRA) strategy was utilized to upgrade the cycle boundaries and primary impacts plot was attracted to notice the effect of boundaries over MRR and surface harshness. It is seen that MRR and surface unpleasantness were straightforwardly corresponding to release current.

Gurrappa, I. [48] The conclusion that the use of titanium alloys is expedient and effective is based on a study of the working conditions and knowledge of the operation of titanium alloys in vessel structures and power plants. When it comes to the application of titanium and its alloys to the development of assemblies or marine equipment, consideration should be given not only to technical performance but also to cost-effectiveness. For the full operating period of the equipment, the cost-effectiveness must be determined, taking into account the savings due to the decline in the cost of repairs, the decrease in the cost of replacement of repair facilities, the decrease in the cost of maintenance of repair services, the decrease in the cost of maintaining the protection of the community and other costs.

He et al. [49] The present study explored the use of side-flow ECM in order to achieve Ti 6Al 4V with a high surface quality, paying particular attention to the relationship between the feed rate and the mean processing current, balance gap, and surface roughness. The surface topographies of titanium alloys, electrochemically machined with varying current densities, were investigated using ECM channel design and analysis. The findings showed that the average processing current increases linearly with the rising feed rate, while the surface roughness decreases rapidly.

Chen et al. [50] In this paper to produce micro-dimple arrays on the surface of Ti-6Al-4V using TMEMM with a PDMS mask, an environmentally friendly $NaNO_3$ electrolyte was used. A pulsed current was employed in TMEMM to obtain high-quality micro-dimple arrays. The results showed that machining parameters of 24 V applied voltage, 10% pulse duty cycle and 100 Hz frequency, were suitable for improving machining efficiency.

Ezugwu et al. [51] This paper deals with the optimization of AISI 202 austenitic stainless steel turning machining parameters using CVD coated cemented carbide materials. Process parameters such as speed, feed, cut depth, and nose radius are used during the experiment to investigate their effect on the work piece's surface roughness (Ra). In the Design of Experiments (DOE) on Computer Numerical Controlled (CNC) lathe, the experiments were performed using a full factorial design. In addition, the study of analysis of variance (ANOVA) was used to analyse the effect and interaction of process parameters during machining.

Mankindan et al. [52] The main objective of this work is to evaluate the operating parameters of Ti-6Al-4V titanium alloy electrochemical drilling. It examines the effect of different parameters such as feed rate, electrolyte flow rate, and electrolyte concentration on the rate of material removal and over cutting. In order to analyse the effect of these parameters, Taguchi's orthogonal arrays, signal-to-noise ratio (S/N ratio), ANOVA and grey relational analysis are used.

Fang et al. [53] In this paper, the hub electrolyte flushing is introduced to WECM for eliminating electrolysis items and recharging electrolyte. Trial results show that WECM with pivotal electrolyte flushing is a promising issue in the manufacture of titanium combination (TC1). The attainability of multi-WECM is additionally shown to improve the machining efficiency of WECM.

4.7 CONCLUSIONS

The principal troubles of ECM of titanium are related with the way that titanium is inclined to passivation in the fluid electrolyte arrangements. Passivation forestalls accomplishing high paces of titanium disintegration that are needed for ECM. This trouble can be forestalled by utilizing arrangements of salts containing anions, which have the initiating impact on aloof titanium. The least capability of anodic–anionic initiation is accomplished in the NaBr arrangements. In this manner, these arrangements are utilized as the electrolytes for ECM independently or in a blend with different salts, for instance, NaCl. The last is oftentimes utilized for ECM of titanium in the individual structure. The aftereffects of ECM of titanium composites rely transcendently upon the titanium content in the amalgam. The strategy for ECM has numerous focal points over the customary techniques for machining and can be utilized in the biomedical and airplane industry and in different fields.

MRR boundary is affected by applied voltage and device feed rate instead of different cycle boundaries. Titanium alloy composites are prevalently utilized in marine designing, clinical and substance ventures in view of its astounding erosion opposition, high explicit quality, and cryogenic properties for optimization. Titanium compound is broadly utilized in different enterprises like aviation and vehicle ventures since it has great solidarity to weight proportion. Material removal rates is quiet high as compared to other machining process due to localization of chemical reactions in small zone by the final optimization techniques. ECM provides the advantage of not affecting the grain structure of the substrate as there is no HAZ.

REFERENCES

1. Davydov AD, Kabanova TB, Volgin VM, Electrochemical machining of titanium. Review. *Russian Journal of Electrochemistry*, 53.9 (2017) 941–965.
2. Anasane SS, Bhattacharyya B, Experimental investigation on suitability of electrolytes for electrochemical micromachining of titanium. *The International Journal of Advanced Manufacturing Technology*, 86.5–8 (2016) 2147–2160.
3. Rao PS, Jain PK, Dwivedi DK, Electro chemical honing (ECH) of external cylindrical surfaces of titanium alloy. *Procedia Engineering*, 100 (2015), 936–945.
4. Sudhakar Rao P, Jain PK, Dwivedi DK, Precision finishing of external cylindrical surfaces of EN8 steel by electro chemical honing (ECH) process using OFAT technique. *Materials Today Proceedings*, 2 (2015) 3220–3229.
5. Bhattacharyya B, Munda J, Malapati M, Advancement in electrochemical micro-machining. *International Journal of Machine Tools and Manufacture*, 44.15 (2004) 1577–1589.
6. Gangwar Arun Kumar Singh, Rao P Sudhakar, Kumar Ashwani, Patil Pravin P, Design and analysis of femur bone: BioMechanical aspects. *Journal of Critical Reviews*, 6.4 (2019) 133–139.
7. Singh T, Misra JP, Upadhyay V, Sudhakar Rao P, An adaptive neuro-fuzzy inference system (ANFIS) for Wire-EDM of ballistic grade aluminium alloy. *International Journal of Automotive and Mechanical Engineering*, 15.2 (2018) 5295–5307.
8. Babar PD, Jadhav BR, Experimental study on parametric optimization of titanium based alloy (Ti-6Al-4V) in electrochemical machining process. *International Journal of Innovations in Engineering and Technology*, 2 (2013) 171–175.

9. Babbar A, Sharma A, Singh P, Multi-objective optimization of magnetic abrasive finishing using grey relational analysis. *Materials Today Proceedings*, 50 (2022) 570–575. https://doi.org/10.1016/j.matpr.2021.01.004

10. Sharma A, Kumar V, Babbar A, Dhawan V, Kotecha K, Experimental investigation and optimization of electric discharge machining process parameters using Grey-fuzzy-based hybrid techniques. *Materials*, 14 (2021) 5820. https://doi.org/10.3390/ma14195820

11. Rao RV, Kalyankar VD, Optimization of modern machining processes using advanced optimization techniques: A review. *The International Journal of Advanced Manufacturing Technology*, 73.5–8 (2014) 1159–1188.

12. Benedict GF *Nontraditional Manufacturing Processes*, Exeter: Marcel Dekker Inc., 1987.

13. Donachie MJ Jr, *Titanium a Technical Guide*, second edition, Ohio: ASM international, 2000.

14. Sharma A, Kalsia M, Uppal AS, Babbar A, Dhawan V, Machining of hard and brittle materials: A comprehensive review. *Materials Today Proceedings* 50 (2022) 1048–1052. https://doi.org/10.1016/j.matpr.2021.07.452

15. Kumar V, Prakash C, Babbar A, Choudhary S, Sharma A, Uppal AS, Additive manufacturing in biomedical engineering. *Additive Manufacturing Processes in Biomedical Engineering* (2022) 143–164. https://doi.org/10.1201/9781003217961-8

16. Acharya BG, Jain VK, Batra JL, Multi-objective optimization of the ECM process. *Precision Engineering* 8.2 (1986) 88–96.

17. Rao PS, Jain PK, Dwivedi DK, Electro chemical honing (ECH) of external cylindrical surfaces—An innovative step. *Daaam International Vienna Publishers*, 09 Chapter, Austria: DAAAM International Scientific Book, 2015, 097–116.

18. Rao PS, Jain PK, Dwivedi DK, Electro chemical honing (ECH) – A new paradigm in hybrid machining process. *DAAAM International Vienna Publishers*, 26 Chapter, Austria: DAAAM International Scientific Book, 2016, 287–306.

19. Sharma A, Jain V, Gupta D, Babbar A, A review study on miniaturization. *Advanced Manufacturing and Processing Technology*, First Edition, Boca Raton: CRC Press, 2020, 111–131. https://doi.org/10.1201/9780429298042-5

20. Babbar A, Jain V, Gupta D, Prakash C, Sharma A, Fabrication and machining methods of composites for aerospace applications. *Characterization, Testing, Measurement, and Metrology*, First Edition, Boca Raton: CRC Press, 2020, 109–124. https://doi.org/10.1201/9780429298073-7

21. Horgan IF, *Electrolytic Boost for Honing*, American Machine, 1962.

22. Asokan P, Kumar RR, Jeyapaul R, Santhi M, Development of multi-objective optimization models for electrochemical machining process. *The International Journal of Advanced Manufacturing Technology*, 39.1–2 (2008) 55–63.

23. Rao RV, Pawar PJ, Shankar Ravi, Multi-objective optimization of electrochemical machining process parameters using a particle swarm optimization algorithm. *Proceedings of the Institution of Mechanical Engineers, Part B: Journal of Engineering Manufacture*, 222.8 (2008) 949–958.

24. Rosenkranz C, Lohrengel MM, Schultze JW, The surface structure during pulsed ECM of iron in $NaNO_3$. *Electrochimica Acta*, 50.10 (2005) 2009–2016.

25. Babbar A, Jain V, Gupta D, Sharma A, Fabrication of microchannels using conventional and hybrid processes. *Non-Conventional Hybrid Machining Processes*, First Edition, Boca Raton: CRC Press, 2020, 37–51. https://doi.org/10.1201/9780429029165-3

26. Sharma A, Grover V, Babbar A, Rani R, A trending nonconventional hybrid finishing/machining process. *Non-Conventional Hybrid Machining Processes*, First Edition, Boca Raton: CRC Press, 2020, 79–93. https://doi.org/10.1201/9780429029165-5

27. Rabbo MF Abd, Boden PJ, Development of electrolytes for the electrochemical machining of titanium I. Electrochemistry in static solutions. *British Corrosion Journal*, 14.4 (1979) 240–245.

28. Ahn Se Hyun, Electro-chemical micro drilling using ultra short pulses. *Precision Engineering*, 28.2 (2004) 129–134.

29. Thakur IS, Pandey VS, Rao PS, Tyagi S, Goyal Deepam, Tribological study of mechanically milled graphite nanoparticles codeposited in electroless Ni-P coatings. *Journal of Metal Powder Report*, 75.6(2020) 344–349.

30. De Silva AKM, Altena HSJ, McGeough JA, Precision ECM by process characteristic modelling. *CIRP Annals*, 49.1 (2000) 151–155.

31. Acharya BG, Jain VK, Batra JL, Multi-objective optimization of the ECM process. *Precision Engineering*, 8.2 (1986) 88–96.

32. Baraiya R, Babbar A, Jain V, Gupta D, In-situ simultaneous surface finishing using abrasive flow machining via novel fixture. *Journal of Manufacturing Processes*, 50 (2020) 266–278. https://doi.org/10.1016/j.jmapro.2019.12.051

33. Uppal AS, Sharma A, Babbar A, Singh K, Singh AK, Minimum quality lubricant (MQL) for ultraprecision machining of titanium nitride-coated carbide inserts: Sustainable Manufacturing process. *International Journal on Interactive Design and Manufacturing (IJIDeM)*, 2023 (2023) 1–12.

34. Babbar A, Jain V, Gupta D, Prakash C, Experimental investigation and parametric optimization of neurosurgical bone grinding under bio-mimic environment. *Surface Review and Letters*, 28 (2021 Jul) 2141005.

35. Singh G, Babbar A, Jain V, Gupta D, Comparative statement for diametric delamination in drilling of cortical bone with conventional and ultrasonic assisted drilling techniques. *Journal of Orthopaedics*, 1.25 (2021 May) 53–58.

36. Gorynin IV, Titanium alloys for marine application. *Materials Science and Engineering: A*, 263.2 (1999) 112–116.

37. Sharma A, Babbar A, Tian Y, Pathri BP, Gupta M, Singh R, Machining of ceramic materials: A state-of-the-art review. *International Journal on Interactive Design and Manufacturing (IJIDeM)*, 2022. https://doi.org/10.1007/s12008-022-01016-7

38. Tamilarasan A, Rajamani D, Multi-response optimization of Nd: YAG laser cutting parameters of Ti-6Al-4V superalloy sheet. *Journal of Mechanical Science and Technology*, 31.2 (2017) 813–821.

39. Khan Mohd Yunus, Rao P Sudhakar, Hybridization of electrical discharge machining process *International Journal of Engineering and Advanced Technology (IJEAT)*, 9.1 (2019) 1060–1065.

40. Khan Mohd Yunus, Sudhakar Rao P, Pabla BS, An experimental study on magnetic field-assisted- EDM process for Inconel-625 *Journal Advances in Materials and Processing Technologies*, 8.sup4 (2022) 2204–2230.

41. Baehre D, Ernst A, Weibhaar K, Natter H, Stolpe, M, Busch, Electrochemical dissolution behavior of titanium and titanium-based alloys in different electrolytes. *Procedia Cirp*, 42.18th (n.d.) 137–142.

42. Chen Xiaolei, Qu Ningsong, Zhibao Hou, Electrochemical micromachining of micro-dimple arrays on the surface of Ti-6Al-4V with $NaNO_3$ electrolyte. *The International Journal of Advanced Manufacturing Technology*, 88.1–4 (2017) 565–574.

43. Liu Jia, Di Zhu, Long Zhao, Xu Zhengyang, Experimental investigation on electrochemical machining of γ-TiAl inter metallic. *Procedia CIRP*, 35 (2015) 20–24.

44. Manikandan N, Kumanan S, Sathiyanarayanan C, Multi response optimization of electrochemical drilling of titanium Ti6Al4V alloy using Taguchi based grey relational analysis. *Engineering Science and Technology, an International Journal*, 20 (2017) 662–71.

45. Fang Xiaolong, Qu Ningsong, Yudong Zhang, Xu Zhengyang, Di Zhu, Effects of pulsating electrolyte flow in electrochemical machining. *Journal of Materials Processing Technology*, 214.1 (2014) 36–43.
46. Weinmann Martin, et al., Electrochemical dissolution behaviour of Ti90Al6V4 and Ti60Al40 used for ECM applications. *Journal of Solid State Electrochemistry*, 19.2 (2015) 485–495.
47. Kumar Ramanuj, Analysis of MRR and surface roughness in machining Ti-6Al-4V ELI titanium alloy using EDM process. *Procedia Manufacturing*, 20 (2018) 358–364.
48. Gurrappa I, Characterization of titanium alloy Ti-6Al-4V for chemical, marine and industrial applications. *Materials Characterization*, 51.2–3 (2003) 131–139.
49. Chen Xiaolei, Qu Ningsong, Zhibao Hou, Electrochemical micromachining of microdimple arrays on the surface of Ti-6Al-4V with NaNO$_3$ electrolyte. *The International Journal of Advanced Manufacturing Technology* 88.1–4 (2017) 565–574.
50. He YF, Electrochemical machining of titanium alloy based on NaCl electrolyte solution. *International Journal of Electrochemical Science* 13.6 (2018) 5736–57477.
51. Ezugwu EO, ZM Wang, Titanium alloys and their machinability—A review. *Journal of Materials Processing Technology* 68.3 (1997) 262–274.
52. Manikandan N, Kumanan S, Sathiyanarayanan C, Multi response optimization of electrochemical drilling of titanium Ti6Al4V alloy using Taguchi based grey relational analysis. (2015).
53. Fang Xiaolong, Effects of pulsating electrolyte flow in electrochemical machining. *Journal of Materials Processing Technology* 214.1 (2014) 36–43.

5 Green Composite
Fabrication, Characterization, Evaluation, and Application

Jimmy Mehta, Yaman Hooda, and Prateek Mittal
Manav Rachna International Institute of Research
and Studies, Faridabad, India

Pallav Gupta
Amity School of Engineering and Technology,
Amity University, Noida, India

5.1 INTRODUCTION

A shift towards the sustainable environment is the need of hour because of change in global climate, glaciers receding, rivers and lakes melting, etc. These issues result in increasing interest in green composites that can combine unique properties and characteristics of metals, polymers and alloys. Sustainability and globalization have made life not only easy feasible but complex too. Materials originally obtained from resources are now renewed for suitable and sustainable development. Varieties of new materials have been explored for applications like furniture, food, textile industry, etc. Recently these materials have gained a lot of popularity and are a hot topic for researchers. These materials act to counter various environmental issues including waste management, global warming, rises in oil prices, deterioration of fossil fuels etc. Composite materials combine two dissimilar materials when integrated together resulting in a new material that has different properties than the parent materials. Composites can be tailored as per the requirements for a particular application and purpose. The mixed parent materials are known as the matrix (major percentage) and the reinforcement (minor percentage). The term green composites are new age composite materials having varied characteristics and a few unique properties that include biodegradability, recyclability, and environmental friendliness. Awareness related to exhausting petrochemical resources, limited availability of resources and reduced carbon emissions are the major reasons for development of green composite materials [1]. Composites and fibres are used as per their acquired properties in varied applications. Animal fibres find their major applications in bio-composite materials and also can be renewed easily. This field is still being explored by researchers but early experiments on plant fibre have given promising results toward developing sustainable materials. In recent times, various studies have been performed to study various machining operations on different kinds of material including composite materials [2–25].

DOI: 10.1201/9781003427735-5

5.1.1 BACKGROUND AND SIGNIFICANCE OF GREEN COMPOSITES

Composite materials posses a large number of advantages and applications, but have a few limitations too. Composite materials that are manufactured have low ductility and temperature limits. As there is more than one material present, composites can be susceptible to moisture or solvents. The matrix component is generally weaker and leads to reduced toughness, high brittleness, risk of damage. and weak transverse properties. Figure 5.1 details the types of matrices that are generally being employed for the manufacture of composite materials.

5.1.2 LIMITATIONS OF COMPOSITE MATERIAL

Despite the numerous advantages of composite materials, their utilization in dynamic structures can be challenging due to their complexity and higher cost compared to conventional materials. Composite materials demand more advanced design, fabrication, and testing techniques, along with specialized skills and equipment for handling and maintenance as their behaviour, properties and characteristics are influenced by multiple variables and interactions, making them difficult to accurately model and analyse. Also these materials have poor wettability, have high moisture absorption and low interface adhesion related to a few polymer matrices. Due to these issues, mechanical characteristics like flexural modulus and fracture toughness decreases. A few chemical treatments like alkali silane, acetic anhydride, combination have

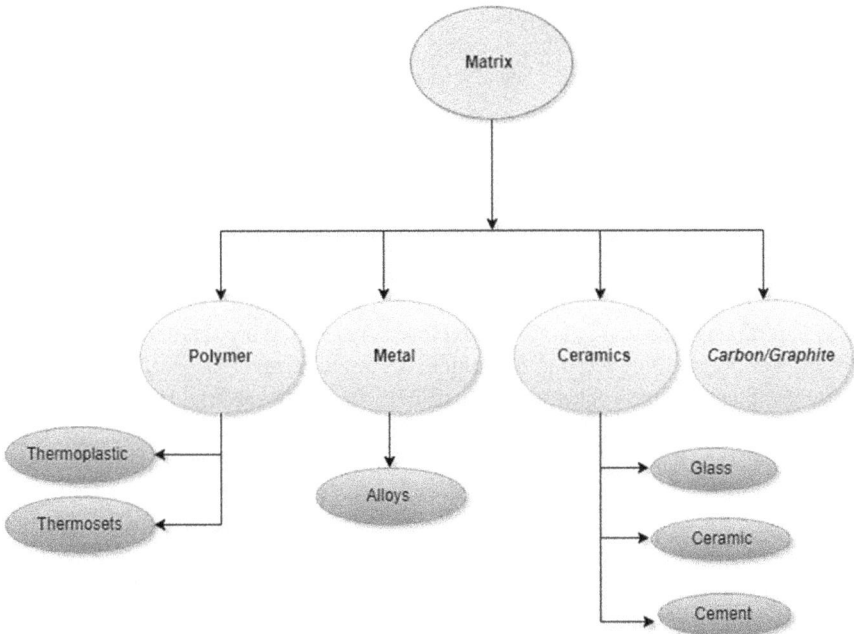

FIGURE 5.1 Types of matrices.

been proposed techniques to reduce this moisture absorption and enhance mechanical properties.

Another aspect is the inherent variability and scatter in the properties and performance of composite materials, which exceeds that of conventional materials. This increased sensitivity to defects, flaws, and environmental factors can pose additional challenges. Moreover, the design and certification of composite materials have less supporting data and experience, leading to higher levels of uncertainty and risk. These require sophisticated approaches for design, fabrication, and testing. Their complex behaviour, higher sensitivity to defects, and limited data availability contribute to the challenges associated with their implementation in dynamic structures. A few disadvantages can be listed as follows:

- Costly material;
- Specialized production processes;
- Necessity for high-quality moulds;
- Cannot be easily recycled;
- Requires proper finishes to encapsulate surface fibres;
- Cannot be easily repaired as structure loses integrity.

5.2 GREEN COMPOSITES AND THEIR ADVANTAGES

Green composites are generally defined as the sustainable materials having matrices that are composed of natural fibres that are economical and an alternate for synthetic fibres. Sustainable word here includes both reactants and bi-products to be harmless and reused or recycled after synthesis [26]. These biomaterials are generally composed of sisal, flax, coir, soybean, hemp, etc. Bio-based polymers include furans, poly-lactic acids, and starch. These bio-reinforced green composites are getting attention in various application fields, products, aviation and automation industry, medical appliances, and packaging and manufacturing to promote more environmental sustainability. Usually green composites are synthesized from agricultural wastes, applicable for both structural and on-structural civil engineering, roofs, retaining, cladding and partition walls, rigid pavements, and embankments. Figure 5.2 enlists major constituents of green composite materials. Association between material characteristics and composite materials that result in durability and recycling with respect to structural enactment is a largely unexplored area and many scholars and researchers are working on it in detail to find quick, economical, and reasonable solutions for resources that can be recycled and reused [27]. Natural fibres used here are the abundantly available, easily accessible, renewable, sustainable, and biodegradable. These have gained popularity in commercial applications involving semi-structural aircraft and automobile parts or sport equipment's electronic appliances. On the other hand, plant fibres are light in weight and high in performance. The main factors affecting the properties and interfacial properties of composites are the main processing parameters. Techniques to improve the strength of natural fibres include chemical treatments and mechanical stress methods. Major benefits of using natural fibres and resin are lower cost, biodegradability and sustainability.

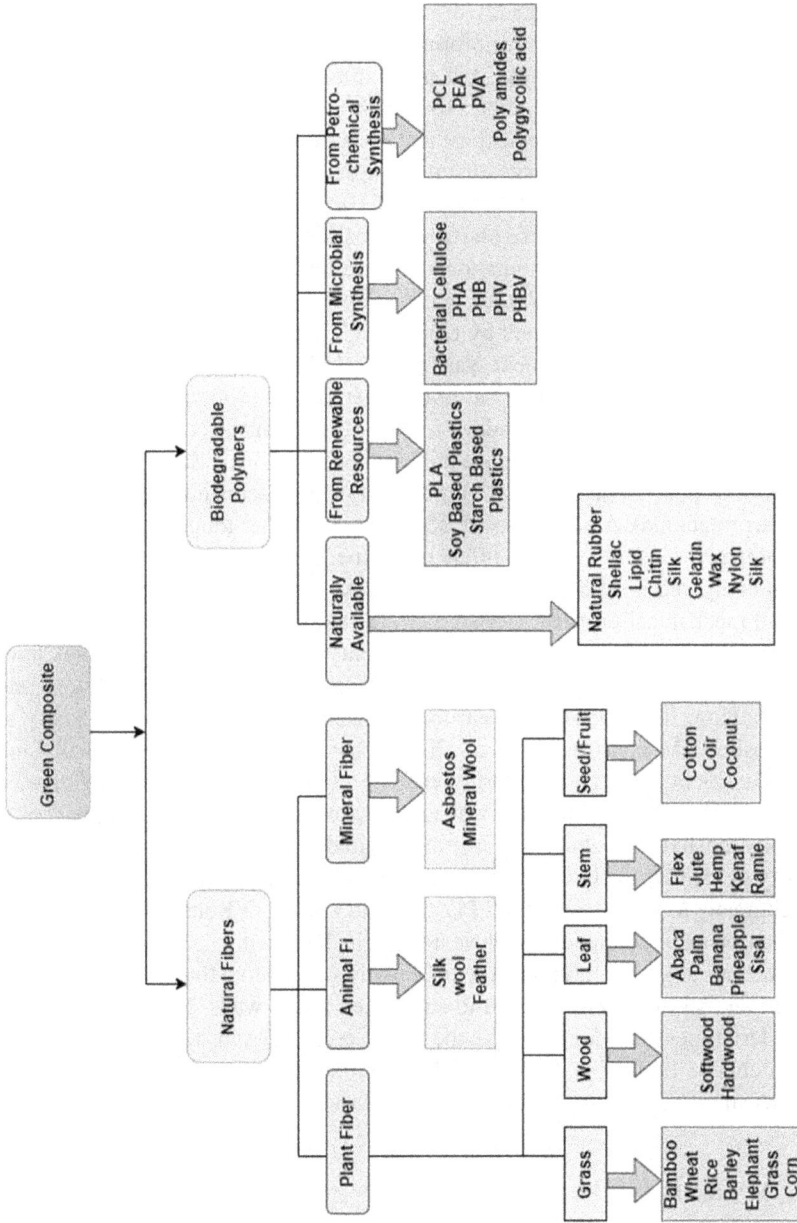

FIGURE 5.2 Constituents of green composite materials.

5.3 PREVIOUS RESEARCH AND DEVELOPMENTS IN GREEN COMPOSITES

Much research has been carried out on characterization and properties of green composites but how are they manufactured is still under research and under progress. Injection moulding, extrusion, thermoforming, and compression moulding are a few named techniques being employed, but not limited, to that. Various authors have studied these manufacturing techniques like Altun et al. [28] who employed injection moulding for manufacturing wood flour and PLA composites, Hu et al. [29] manufactured jute and PLA composites using compression moulding, whereas Chang et al. [30] used thermoforming technique and Faruk et al. [31] reported usage of RTM. Increasing demand of these plastic composite materials that have substituted other materials have led to combinations of these technologies, for example direct long fibre thermoplastics (DLFTs). Increasing interest for these composite materials has been generated in recent years by employing plant-based natural fibres instead of synthetic fibres for reinforcement. Varied natural fibres have been utilized, including banana, aloe vera, kenaf, and sisal fibres. Mechanical properties have also been explored by different scholars that added rice husk, wheat husk, coconut coir as natural fibre ingredients. Coconut coir gave the best result in GFRP. Biodegradable polymers have like poly lactic acid (PLA), poly butylene succinate (PBS), and maybe poly (e-carprolactone). Among these, PBS has shown better toughness, and elongation as compared to PLA, but PLA offers better mechanical and thermal characteristics when compared to PBS. Also, animal fibre like worm silk and wool have better elastic and mechanical properties. Also, uniform topography is its major advantage as compared to plant based natural fibres. Various manufacturing techniques have been developed, such as winding, centrifugal casting (for tubes), pultrusion, and vacuum shaping. In the 1990s, infrastructure construction was the first industry to use composites. The construction of an all-composite bridge at Aberfeldy Golf Club in Perthshire, Scotland, is a noteworthy example. Given that the River Tay divided the area, the bridge's main function was to increase the size of the nine-hole course that already existed. The club's administrators hired Prof. Bill Harvey from Dundee University to identify the best answer. Prof. Harvey worked with senior students, the design company Maunsell Structural Plastics, and builder O'Rourke. The first green composites were made hundreds of years ago in Mesopotamia by joining timber strips at various angles to form the precursor to plywood. More than a thousand years later, mud and straw were employed to strengthen house walls. Moving forward to more modern times, the Mongols were the first to create composite bows, which used wood, bamboo, silk, bovine tendons, and glue from pine trees. The arrow shooting power was significantly boosted by a factor of ten with these composite bows.

5.3.1 MATERIALS AND METHODS

While the twentieth century was characterized by the predominance of concrete, the nineteenth century is frequently referred to as the era dominated by steel. However, with the widespread usage of polyester-glass composites in the twenty-first century, there has been a resurgence in interest in wood as a building material.

The fact that wood is the only naturally occurring and renewable material that is widely used in everyday construction is one of the key factors in this resurgence in popularity. Matrix and dispersion are the two separate phases of composite material. Manufactured materials like minerals, fibres (platelets or particles), or natural or other synthesized materials are all included in the dispersal phase. The matrix phase can be natural or synthetic polymer. Intermittent surface helps in analysing the reinforcement efficiency of composite material prepared and this bonding is essential for all the characteristics and properties to be measured. Wettability between the mating surfaces has to be suitable.

5.4 SELECTION OF RAW MATERIALS AND FABRICATION TECHNIQUES

Surface techniques can also be used to improve the properties. A petroleum based non-biodegradable polymers matrix results in partially biodegradable polymers. Green composites can be either discontinuous reinforced composites or unidirectional and bidirectional continuous composites, depending on the type of reinforcement used. Additionally, aligned or randomly oriented particles, short fibres, and whiskers are categorized as discontinuous composites. Few scholars have talked about the three main methods for creating green composites. These are as discussed in Figure 5.3.

Green composites made from plant sources that are biodegradable with high strength and stiffness are being developed with potential applications in structures and construction materials.

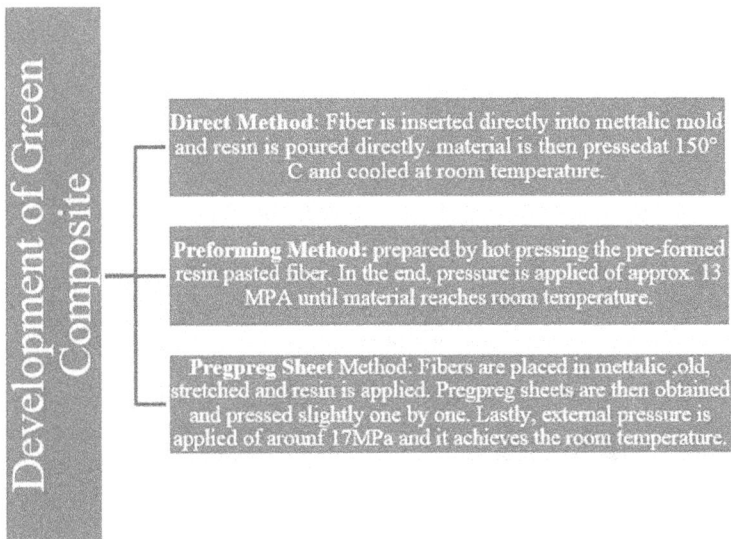

FIGURE 5.3 Development of green composite.

In compression Moulding, the moulding material is preheated and placed in the cavity. The mould is closed by pressing the top until it has solidified. Thermosetting resins are used either as granules, mastic masses or voids. It is a high-volume, high-pressure method, so it is best suited for complex moulded parts and strong fibreglass reinforcements. It is one of the most economical methods and has almost negligible wastage of material providing an advantage, particularly when working with expensive materials. The basic aim of developing this technique was to make big parts flat or slightly curved. Thermoplastic and thermoset materials are melted in a mould cavity at a high temperature and pressure, then immediately cooled, the post-process component is removed, and the substance solidifies. Chen [32] studied the fabrication of bio-composite materials that were prepared using recycled polymers and rice husk fibres. The fibres were added in 40–80% weight ratio, and tensile strength was checked along with water absorption and thermal absorptivity. Additionally, Ahmed and Sivaganesan [33] investigated the effects of altering the proportions of neem and glass hybrid fibres (10–40%) and rice husk filler (2% by weight) when making a green epoxy composite material using compression moulding. They experimented for tribological behaviour using a pin-on-disk tribometer. Many other scholars and researchers have also studied and experimented with epoxy bio-composite material when added with varying percentages of neem fibre, glass fibre, and rice husk [34–36]. Also, Figure 5.4 details variouslfabrication techniques being used for green composite materials. A few important points to be considered before employing any technique are as follows:

- Volume of material
- Amount of energy needed to heat that substance
- Heat required by the material
- Process/Technique used for heating
- Force and pressure required to attain required shape
- Rapid cooling system after material has been compressed.

Raw material selection starts at an early stage for the development of a new product/material. Selection criteria depend on cost, properties required, and process ability of green composites. Low cost and processing temperatures have resulted in benefits for manufacturers. Properties to consider include mechanical, physical, chemical and thermal properties. It is the first and foremost stage where the work begins and when end result that is the application part has been decided where material would be used. Table 5.1 details the major difference points between compression moulding (Figure 5.5) and thermoforming techniques.

1. Injection Moulding: One of the most common technique for producing parts by plastics. In this process, the heated material is injected into mould by applying external force. During solidification and cooling, the part is ejected and process is continued as shown in Figure 5.6. This method helps in producing variety of customized materials using varied raw material in just one single operation. Thermoset plastics when used have to undergo

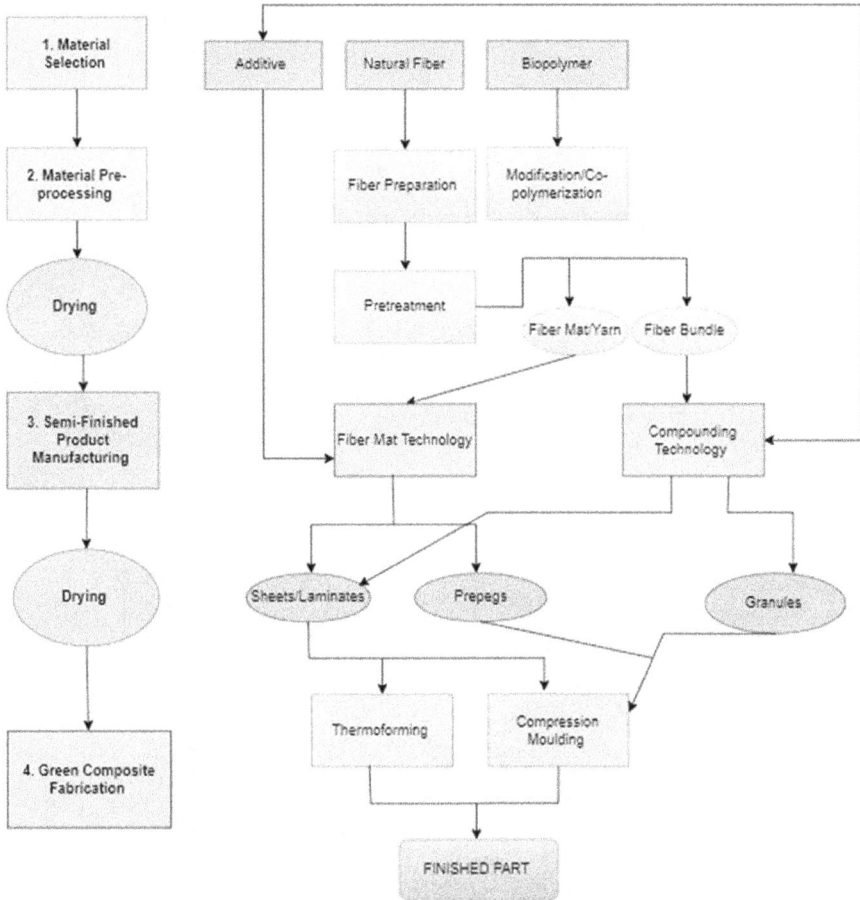

FIGURE 5.4 Flowchart for the fabrication of green composite.

TABLE 5.1
Difference between Compression Moulding and Thermoforming Techniques

S.No.	Features	Compression Moulding	Thermoforming
1	Type of Process	Discontinuous	Continuous
2	Volume and tolerance	Low volume production with high tolerance parts	Great volume production with low tolerance
3	Availability of Raw Material	Sheets, Composite Granules, Prepegs	Sheets
4	Material Waste	Low	High
5	Application	Automotive hood, fender, door panels, gears etc	Food Packaging, aircraft windscreen, interior, panels, etc

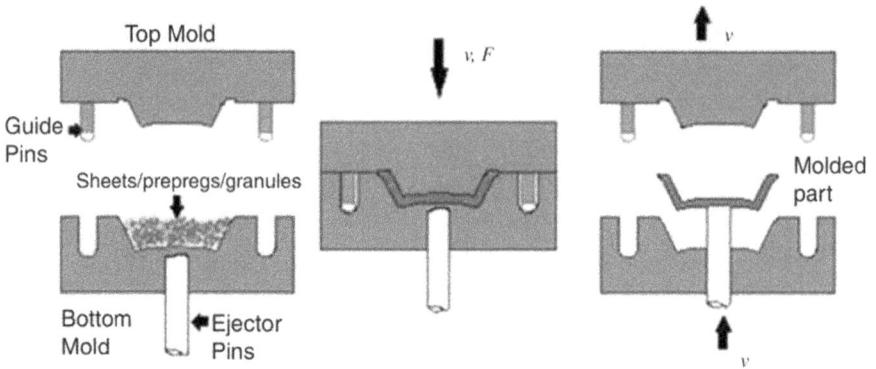

FIGURE 5.5   Compression moulding apparatus.

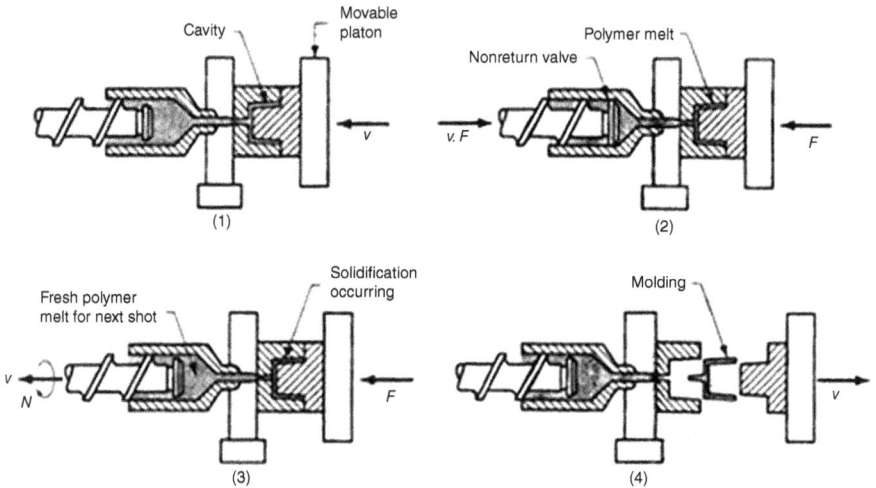

FIGURE 5.6   Injection moulding set up.

chemical reactions and once formed into shape can't be deformed, hence thermoplastics are preferred. Major influencing factors are as follows:

- Molder
- Material
- Injection machine apparatus
- Mold

Among these, apparatus and mould vary in design and assembly. This machine gives advantage of high production output rates with close tolerances. Figure 5.6 details the setup of injection moulding process.

Green composites find their major applications in panels employed for automobiles as these can be easily compared with the ones prepared by synthetic materials based on mechanical properties [37]. However, because they are desiged to decompose easily, these green composites cannot be utilized for 100% bio-based composites when considered for exterior panels for vehicles. Researchers have experimented and successfully replaced fibreglass in strengthened plastic composites by common plant strands like flax, jute, sisal, ramie, etc. [38–41]. An endeavour was made in 1996 for Mercedes-Benz by utilizing epoxy framework with expansion of jute in entryway boards in E-class vehicles [42]. Another endeavour was made in 2000 in Audi A2 midrange cars as entryway boards were made of polyurethane fortified with flax and sisal [20]. Also in this league Toyota leads the competition by using environmentally friendly materials, i.e., 100% bioplastics. Spare tire cover parts were manufactured using PLA matrix (sugarcane, sweet potato) reinforced with kenaf fibre [43]. Mitsubishi motors, Toyota, Ford, BMW and many others are widely using composite material for various parts of automobiles and experimenting for the reinforced material as well as matrix [44–47]. From a study, it has been concluded that car weight has been reduced up to 35%. These biodegradable plastics have earned a great market and attention as they have ecological sustainability. Number of applications in all diverse fields are found and more research is still going on striving for improved environmental friendliness, sustainability, reusability, and recyclability. Slight variation in composition and percentage of material added, application and use of material varies and every component has its own pros and cons.

5.5 CHARACTERIZATION TECHNIQUES FOR EVALUATING THE PROPERTIES OF GREEN COMPOSITES

5.5.1 Mechanical Properties Analysis

Natural fibres naturally have poorer mechanical qualities than glass fibres, hence green composites that use them as reinforcement frequently have worse mechanical characteristics than glass fibre-reinforced composites. The general characteristics of green composites can also be influenced by other characteristics like hydrophilicity, water absorption, heat deterioration of the fibres, poor fibre/matrix adhesion, and others. Optimizing the fibre/matrix contact is essential for overcoming these problems and improving the mechanical properties. To achieve this optimization, physical and chemical approaches can both be used. Stretching, heat treatment, plasma treatment, and electric discharge are examples of physical and mechanical techniques that can influence the structural and surface properties of fibres without changing their chemical makeup, having an impact on the mechanical properties of composites. Natural fibres, however, are fundamentally incompatible with hydrophobic resins due to their strong polarization. To increase fibre/matrix adhesion utilizing coupling agents in composites, chemical techniques such adjusting the fibre surface pressure, chemical coupling, acetylation, and soluble base treatment are thought to be beneficial. Natural fibres treated with alkali are one way to increase the toughness of green composites. When ramie fibres were subjected to an alkali treatment with a 15% NaOH solution and a load, the fibres' fracture strain increased dramatically without noticeably affecting their tensile strength. This increase in toughness is directly related to the reduction in fracture strain. Although the concentration utilized in the study was too low to completely convert all of the cellulose I to cellulose II, the alkali treatment changes the crystal structure of cellulose I to cellulose II. In addition, the stress applied during the alkali treatment may reduce the micro-fibrils' angles, increasing the fibre strength. When alkali-treated ramie plied yarns were used to create green composites, the composites showed a slight loss in tensile strength and a significant rise in fracture strain, leading to noticeably higher toughness. Compared to untreated composites, these composites' impact resistance was dramatically increased. In comparison to glass fibre-reinforced composites, textile green composites laminated using alkali-treated ramie woven fabrics also shown higher impact resistance. When a composite's mechanical properties were examined, it was discovered that the tensile strength, flexural strength, and impact strength had all decreased by 7.69%, 12.06%, and 3.27%, respectively. Due to their exceptional qualities, cost-effectiveness, and adaptability for a variety of applications, organic fibre-reinforced polymer hybrids (NFPC) have become increasingly popular in the transportation industry. However, these fibres have shortcomings such as high hydrophilicity and low dimensional stability. The physical behaviour of the composite is greatly influenced by the interfacial bonding between the reinforcement and matrix. Different chemical treatments are used to promote the adherence of the fibres to the matrix, leading to composites with superior mechanical properties [48]. In order to reduce total weight and improve performance, composites are being thoroughly explored as prospective replacements for conventional high-density materials in the transportation and aviation sectors. This in-depth introduction presents bio-composites, examines the variety of uses for them, and emphasizes the variety of chemical alterations that may be made to improve their qualities.

5.5.2 THERMAL PROPERTIES ANALYSIS

Thermal properties analysis is a crucial aspect when evaluating green composite materials, particularly in terms of thermal stability and heat resistance. Green composites, which utilize environmentally friendly and sustainable reinforcements, require a thorough understanding of their thermal behaviour. Thermal stability analysis assesses the composite's ability to retain its structural integrity and mechanical properties under elevated temperatures, ensuring its suitability for various applications. Heat resistance analysis focuses on determining the material's capacity to withstand high temperatures without significant degradation or dimensional changes. These analyses aid in the selection and optimization of green composite materials for applications where thermal performance is critical, such as in aerospace, automotive, and building industries. The thermal characteristics of composites are crucial as they impact suitable applications and processing methods, particularly in natural fibre and synthetic polymer-matrix composition of fibre. Processing temperatures of the matrix can exceed fibre degradation temperatures. Due to their lower processing temperatures, bio-based polymers are less worrisome in this regard. For fibres, cellulose breakdown happens between 300 and 400°C, hemicellulose decomposition happens at 200°C, and lignin decomposition happens between 400 and 500°C. The temperature range needs to be taken care of with respect to certain polymers for successful synthesis of composite materials.

5.5.3 Morphological Analysis

The SEM/EDX technique is valuable for research involving the identification of elements, whether they originate from within or outside the body, in various biological components such as tissues and cells. This technique is particularly useful in studying drug delivery systems. EDX analysis aids in detecting nanoparticles that are utilized to enhance the therapeutic effectiveness of specific chemotherapy drugs. Additionally, SEM/EDX is employed in environmental pollution studies and for characterizing minerals that accumulate in tissues [49]. For instance, SEM allows for the examination of intricate nano porous aerogel structures whereas in situ SEM strategies explore the warm soundness of nanoparticles like graphene/Cu-based materials, the effect of electron bar illumination on the electrical properties of carbon nanotube yarns, and the nano space properties of multiphase thermoelectric materials [49]. On the other hand, AFM is utilized to explore the estimate and shape of nanoparticles in a three-dimensional mode, decide the degree of surface scope with nanoparticles, analyse nanoparticle scattering in cells and other carriers or lattices, and degree the exact sidelong measurements of nanoparticles [50].

5.5.4 Water Assimilation and Corruption Examination

When managing with natural-fibre composites, the assimilation of dampness plays a significant role in determining the mechanical properties of the composite [50–52]. Given that bio-based polymers have a tendency to absorb more moisture than oil-based polymers do, this factor becomes even more significant when the matrix is made of bio-based polymers. It is crucial to take this restriction into account during the design phase for particular applications since the performance of green composites is negatively impacted when they come into contact with water. The inherent hydrophilicity of plant fibres has a significant impact on moisture absorption in green composites. In addition, fibre volume ratio and ambient temperature were observed to have a significant impact on water absorption, with higher fibre content and temperature leading to increased absorption. The incompatibility between the hydrophobic polymer matrix and the hydrophilic fibre can lead to the formation of micro-pores, which improves water retention [53–55]. By making it easier for water molecules to diffuse inward, the tiny spaces between the polymer's molecular chains also aid in water absorption. The fibre-matrix interface is greatly impacted by the presence of adsorbed water molecules. The adhesion between the fibre and the matrix is reduced as a result of the capillary effect which moves water across the contact [56]. This phenomenon creates areas of low transmission efficiency, weakens the mechanical properties of the composite, causes deformation, and can even lead to degradation.

5.6 EVALUATION OF GREEN COMPOSITES

5.6.1 Performance Comparison with Conventional Composites

The turn of the twenty-first century witnessed significant technological advancements in the field of green composite materials, revolutionizing various aspects of our lives. In a wide range of industries, such as civil engineering, infrastructure, pipeline, and tank construction, offshore work, aerospace, yachts and boats, and sanitary facilities, among others, these have found considerable application as building materials. Their remarkable attributes include lighter weight compared to traditional materials and superior strength. For instance, epoxy resin-based carbon composites can be five times stronger than steel while weighing only 20% of its weight. This makes composites very popular as a construction material. Additionally, composites offer protection against material corrosion, display good chemical and thermal resistance, work well as insulators, and have special features not typically present in traditional materials. In addition to their versatility, composites are easily mouldable, durable, and possess high impact strength. They also offer greater design flexibility and often come at a lower cost compared to certain metals. Overall, the advent of composites has brought about a transformative impact, providing exceptional building materials with a wide range of applications. Their lightweight nature, exceptional strength, resistance to corrosion, and unique properties make them highly desirable across various industries. Additionally, their ease of shaping, durability, high impact strength, design freedom, and cost-effectiveness further contribute to their widespread utilization.

5.6.2 Discussion of the Strengths and Limitations of Green Composites

Unfortunately, even so-called "green" composites are not fully eco-compatible, as their recyclability is limited due to degradation phenomena and temperature constraints (recycling temperatures cannot exceed 200°C). Additionally, the biodegradability of these composites typically only applies to the filler materials, not the petroleum-based polymer matrices used in their composition. To address these limitations, recent research focuses on developing 100% "green" and sustainable composites by replacing non-biodegradable polymeric substrates with biodegradable alternatives. There are several biodegradable polymers derived from natural sources that can be used, such as polysaccharides (starch, chitin, collagen, gelatin), proteins (casein, albumin, silk, elastin), polyesters (e.g. poly(hydroxyalkanoate), poly(hydroxybutyrate) polylactic acid), lignin, lipids, natural rubber, polyamides, polyvinyl alcohol, polyvinyl acetate, and polycaprolactone. These biodegradable polymers usually break down through enzymatic reactions in the right environment, often in the presence of humidity. Examples of eco-sustainable composites can be found in the literature. For instance, Japanese researchers have studied composites made of bamboo and starch fibres. Other investigations have looked into composites made from jute fibres, ananas, and Monsanto Biopol® (a polyhydroxy butyrate-hydroxyvalerate copolymer). Natural fibres have occasionally been pre-treated via chemical group grafting and alkali treatment. Additionally, soy proteins have been used as matrix materials in conjunction with other natural fibres to create composites that are promising and demonstrate qualities that are superior to several types of wood. There have been initiatives to create composites for use in automobiles utilizing polymer matrices based on soy and maize oil, which are used as raw ingredients in the production of polymers. These composites are useful for applications such as panels andseats as they exhibit characteristics including resistance, flexibility, lightness, and durability. Furthermore, some companies have explored the production of synthetic silk through genetic engineering, which could offer possibilities for the development of biodegradable materials [57–63]. Overall, current research has focused on creating 100% environmentally sustainable composites by substituting biodegradable materials generated from natural sources with non-biodegradable polymer matrices. These efforts aim to address the limitations of recyclability and biodegradability in current "green" composites and pave the way for more environmentally friendly materials and applications.

5.7   APPLICATIONS OF GREEN COMPOSITES

5.7.1   CONSTRUCTION

Construction materials have a significant negative environmental impact, with a substantial portion of landfill volume in the U.S. consisting of destruction wastes, including timber, drywall, and plastic. Composites find application in structural and non-structural applications has been explored, considering their physical and chemical characteristics. Researchers have looked into renewable resources made of hemp fibres and poly-hydroxybutyrate, which have mechanical qualities similar to structural wood. This creates opportunities for their application in temporary constructions such as walls, flooring, scaffolding, and formwork. Natural fibres with specific qualities suited for the building sector, such as cotton, coconut, sisal, banana, hemp, sugarcane, and sisal are frequently used as reinforcement in polymer systems.

5.7.2   AUTOMOTIVE INDUSTRY

Green composites have gained traction in the automotive industry for both internal and external components, driven by environmental concerns and European regulations. By 2015, the European Commission must achieve a high level of car recycling in accordance with Directive 2000/53/EG. Dashboards, door panels, seat cushions, and linings are just a few of the automotive body sections that use green composites, which are created from renewable materials. According to experts, an all-advanced-composite auto body might weigh between 50 and 70% less than a steel auto body, while optimized steel vehicle bodies can reduce mass by between 40 and 55 percent. To maximize battery performance, lightweight materials are especially important in electric vehicles. Natural-fibre composites have been included into automotive parts including door panels, interior linings, and seat cushions by renowned automakers like Audi, BMW, Fiat, SEAT, and Volkswagen.

5.7.3   BIOMEDICAL

Tissue engineering is a very active area that concentrates on using novel materials that are compatible with and biodegradable, such poly-D-lactic acid and PLA (poly-lactic acid). Excellent mechanical and biodegradability characteristics can be found in these materials. These polymers have been successfully combined with other biopolymers such chitosan, poly-lactic-co-glycolic acid, and PEG to produce materials with better mechanical properties. Traditional materials for orthopaedic implants or medical purposes include metals like titanium, cobalt-chromium alloys, and stainless steel, as well as polymers like polyethylene and ultra-high molecular weight poly (ether-ethers), ceramics like hydroxyapatite, and polymers like polyethylene and ultra-high molecular weight polyethylene. However, there has recently been a move away from using bio-stable, biocompatible materials in medical devices in favour of bio-absorbable or biodegradable materials to repair and regenerate damaged tissues.

5.8   CONCLUSION

The utilization of polymer composites filled with natural-organic fillers, instead of mineral-inorganic fillers, is a subject of considerable interest due to the reduction in the use of nonrenewable, petroleum-based resources and the efficient consumption of ecological and economic resources. These "green" composites have diverse industrial usage, but limitations arise concerning ductility, processability, and dimensional stability. Globally, there is a lot of research being done to find answers by changing the chemical composition of fillers and using additives and adhesion promoters. By using biodegradable polymers instead of conventional ones, full biodegradability and improved environmental effect can be achieved, but this creates new difficulties. Choosing appropriate biodegradable matrices and optimizing preparation and processing parameters are the main research interests. Particularly in Europe, the market for these composites is still developing, providing chances to expand usage, improve marketability, and discover new applications. The development of formulations using virgin or recycled polymers, conventional or biodegradable polymers, appropriate fillers, characterizing them, and improving processing methods all require major research efforts. As the market expands, cost reductions and quality improvements will be realized. The current global environmental situation necessitates eco-friendly solutions in various sectors, including materials. Green composites (GC) developed from green polymer-based composites and natural reinforces from renewable sources, have gained attention and popularity. These bio-based polymer matrix composites exhibit attractive characteristics, particularly biodegradability, which positions them as environmentally friendly materials in the market. Engineering processes for these resources have seen significant improvements due to the diverse properties of raw materials and their eco-friendly nature, enabling their integration into sectors like construction, automotive, packaging, and medicine. This segment represents a significant economic opportunity, particularly in regions rich in these resources, fostering the development of green economies and advancing global sustainability efforts.

5.9   RECOMMENDATIONS FOR FUTURE RESEARCH

Composites are widely used in various industries due to their lightweight, durability, and customizable properties. Car bumpers made of composites are already used in the automobile sector and the material shows potential for applications in suspensions and brakes. In earthquake-prone areas, cement composites reinforced with fibres offer an alternative to traditional reinforced concrete frames in buildings. Bio-composites, made from natural fibres and oils, are a more sustainable option compared to conventional composites. They are particularly significant in reducing reliance on fossil fuels, as petroleum and coal reserves are limited. Bagasse fibre-based biocomposites are believed to revolutionize future lifestyles due to their availability and low cost. They have promising applications in airplane and vehicle interiors, for example,in shelves, seat covers, and cabin linings. While bio-composites may not completely replace conventional composites in the near future, companies are already engaged in launching eco-friendly and sustainable products. The advantages of bio composites are their light nature, absence of volatile organic compounds and biodegradability. The bio-composite industry is projected to surpass USD 40 million by 2025, with significant growth expected in the construction and manufacturing sectors. Latestly, green composite materials are now being used in vehicle accessories and equipment's, automotive and aerospace industry for development of recyclable and sustainable materials, reuse and safety of building materials. Research is being conducted to enhance the material and manufacturing properties of bio-composites through various methods such as developing new fibre types, utilizing solvent spinning processes for liquid crystalline cellulose, and modifying triglycerides and oils to create cost-effective and biodegradable resins. The aim is to replace synthetic complexes with these environmentally friendly alternatives.

REFERENCES

1. Das R, Bhattacharjee C. Green composites, the next-generation sustainable composite materials: Specific features and applications. *Green Sustainable Process for Chemical and Environmental Engineering and Science*. 2022 Jan; 1: 55–70.
2. Sharma A, Kumar V, Babbar A, Dhawan V, Kotecha K, Prakash C. Experimental investigation and optimization of electric discharge machining process parameters using Grey-fuzzy-based hybrid techniques. *Materials*. 2021 Jan; 14(19): 5820.
3. Prakash C, Kumar V, Mistri A, Uppal AS, Babbar A, Pathri BP, Mago J, Sharma A, Singh S, Wu LY, Zheng HY. Investigation of functionally graded adherents on failure of socket joint of FRP composite tubes. *Materials*. 2021 Jan; 14(21):6365.
4. Babbar A, Jain V, Gupta D, Prakash C. Experimental investigation and parametric optimization of neurosurgical bone grinding under bio-mimic environment. *Surface Review and Letters*. 2021 Jul; 28: 2141005.
5. Babbar A, Jain V, Gupta D, Agrawal D, Prakash C, Singh S, Wu LY, Zheng HY, Królczyk G, Bogdan-Chudy M. Experimental analysis of wear and multi-shape burr loading during neurosurgical bone grinding. *Journal of Materials Research and Technology*. 2021 Feb 24; 12: 15–28.
6. Babbar A, Jain V, Gupta D, Agrawal D. Histological evaluation of thermal damage to Osteocytes: A comparative study of conventional and ultrasonic-assisted bone grinding. *Medical Engineering & Physics*. 2021 Feb 16; 90: 1–8.
7. Babbar A, Jain V, Gupta D, Agrawal D. Finite element simulation and integration of CEM43 °C and Arrhenius Models for ultrasonic-assisted skull bone grinding: A thermal dose model. *Medical Engineering & Physics*. 2021 Feb 16; 90: 9–22.
8. Babbar A, Prakash C, Singh S, Gupta MK, Mia M, Pruncu CI. Application of hybrid nature-inspired algorithm: Single and bi-objective constrained optimization of magnetic abrasive finishing process parameters. *Journal of Materials Research and Technology*. 2020 Jul 1; 9(4): 7961–74.
9. Baraiya R, Babbar A, Jain V, Gupta D. In-situ simultaneous surface finishing using abrasive flow machining via novel fixture. *Journal of Manufacturing Processes*. 2020 Feb 1; 50:266–78.

10. Singh S, Prakash C, Pramanik A, Basak A, Shabadi R, Królczyk G, Bogdan-Chudy M, Babbar A. Magneto-rheological fluid assisted abrasive nanofinishing of β-phase Ti-Nb-Ta-Zr alloy: Parametric appraisal and corrosion analysis. *Materials*. 2020 Jan; 13(22): 5156.

11. Babbar A, Jain V, Gupta D. Preliminary investigations of rotary ultrasonic neurosurgical bone grinding using Grey-Taguchi optimization methodology. *Grey Systems: Theory and Application*. 2020 Jun 23; 10: 479–93.

12. Babbar A, Jain V, Gupta D. In vivo evaluation of machining forces, torque, and bone quality during skull bone grinding. *Proceedings of the Institution of Mechanical Engineers, Part H: Journal of Engineering in Medicine*. 2020 Mar 17; 264:626–38.

13. Babbar A, Jain V, Gupta D. Thermogenesis mitigation using ultrasonic actuation during bone grinding: A hybrid approach using CEM43°C and Arrhenius model. *Journal of the Brazilian Society of Mechanical Sciences and Engineering*. 2019 Oct 1; 41(10): 401. (SCI IF = 1.755)

14. Babbar A, Sharma A, Jain V, Jain AK. Rotary ultrasonic milling of C/SiC composites fabricated using chemical vapor infiltration and needling technique. *Materials Research Express*. 2019 Apr 24. https://doi.org/10.1088/2053-1591/ab1bf7. (SCI IF = 1.929) ISSN: 2053-1591, 08-05-2019.

15. Singh G, Babbar A, Jain V, Gupta D. Comparative statement for diametric delamination in drilling of cortical bone with conventional and ultrasonic assisted drilling techniques. *Journal of Orthopaedics*. 2021 May 1; 25: 53–58.

16. Babbar A, Sharma A, Singh P. Multi-objective optimization of magnetic abrasive finishing using grey relational analysis. *Materials Today: Proceedings*. 2022 Jan 1; 50: 570–5.

17. Sharma A, Kalsia M, Uppal AS, Babbar A, Dhawan V. Machining of hard and brittle materials: A comprehensive review. *Materials Today: Proceedings*. 2022 Jan 1; 50:1048–52.

18. Babbar A, Sharma A, Bansal S, Mago J, Toor V. Potential applications of three-dimensional printing for anatomical simulations and surgical planning. *Materials Today: Proceedings*. 2020 Jan 1; 33: 1558–61.

19. Babbar A, Jain V, Gupta D. Thermo-mechanical aspects and temperature measurement techniques of bone grinding. *Materials Today: Proceedings*. 2020 Jan 1; 33: 1458–62.

20. Sharma A, Babbar A, Jain V, Gupta D. Enhancement of surface roughness for brittle material during rotary ultrasonic machining. In MATEC Web of Conferences 2018 (Vol. 249, p. 01006). EDP Sciences.

21. Babbar A, Singh P, Farwaha HS. Regression model and optimization of magnetic abrasive finishing of flat brass plate. *Indian Journal of Science and Technology*. 2017 Aug; 10:1–7.

22. Sharma A, Jain V, Gupta D. Characterization of chipping and tool wear during drilling of float glass using rotary ultrasonic machining. *Measurement*. 2018 Nov 1; 128: 254–63.

23. Sharma A, Jain V, Gupta D. Comparative analysis of chipping mechanics of float glass during rotary ultrasonic drilling and conventional drilling: For multi-shaped tools. *Machining Science and Technology*. 2019 Jul 4; 23(4):547–68.

24. Bijlwan PP, Prasad L, Sharma A. Recent advancement in the fabrication and characterization of natural fiber reinforced composite: A review. *Materials Today: Proceedings*. 2021 Jan 1; 44: 1718–22.

25. Sharma A, Jain V, Gupta D. Effect of pre and post tempering on hole quality of float glass specimen: For rotary ultrasonic and conventional drilling. *Silicon*. 2021 Jun; 13: 2029–39.

26. Shekar HS, Ramachandra MJ. Green composites: A review. *Materials Today: Proceedings*. 2018 Jan 1; 5(1): 2518–26.

27. KC B, Pervaiz M, Faruk O, Tjong J, Sain M. (2015). Green composite manufacturing via compression molding and thermoforming. In: *Manufacturing of Natural Fibre Reinforced Polymer Composites* (pp. 45–63). New York, USA: Springer International Publishing. https://doi.org/10.1007/978-3-319-07944-8_3

28. Altun Y, Dogan M, Bayramli E. Effect of alkaline treatment and pre-impregnation on mechanical and water absorption properties of pine wood flour containing poly (lactic acid)based green-composites. *Journal of Polymers and the Environment*. 2013; 21: 850–56.

29. Hu R-H, Ma Z-G, Zheng S, Li Y-N, Yang G-H, Kim H-K, Lim J-K. A fabrication process of high volume fraction of jute fiber/polylactide composites for truck liner. *International Journal of Precision Engineering and Manufacturing*. 2012; 13: 1243–6.

30. Chang PR, Zhou Z, Xu P, Chen Y, Zhou S, Huang J. Thermoforming starch-graft-polycaprolactone bio-composites via one-pot microwave assisted ring opening polymerization. *Journal of Applied Polymer Science*. 2009; 113: 2973–9.

31. Faruk O, Bledzki AK, Fink HP, Sain M. Bio-composites reinforced with natural fibers: 2000–2010. *Progress in Polymer Science*. 2012; 37: 1552–96.

32. Chen RS, Ab Ghani MH, Ahmad S, Tarawneh MA, Gan S. Tensile, thermal degradation and water diffusion behaviour of gamma-radiation induced recycled polymer blend/rice husk composites: Experimental and statistical analysis, *Composites Science and Technology*. 2020; 207: 108748.

33. Ahmed AF, Sivaganesan, S. Characterization of material properties of green polymer composite. *Materials Today: Proceedings*. 2022; 69: 789–92. https://doi.org/10.1016/j.matpr.2022.07.213

34. Kumaraswamy J, Kumar V, Purushotham G. A review on mechanical and wear properties of ASTM a 494 M grade nickel-based alloy metal matrix composites. *Materials Today: Proceedings*. 2021; 37: 2027–32.

35. Jayappa K, Kumar V, Purushotham GG. Effect of reinforcements on mechanical properties of nickel alloy hybrid metal matrix composites processed by sand mold technique. *Applied Science and Engineering Progress*. 2020; 14(1): 44–51.

36. Kumaraswamy J, Vijaya Kumar. Evaluation of the microstructure and thermal properties of (ASTM A 494 M grade) nickel alloy hybrid metal matrix composites processed by sand mold casting. *International Journal of Ambient Energy*. 2021; 42: 1–10.

37. Mann GS, Azum N, Khan A, Rub MA, Hassan MI, Fatima K, Asiri AM. Green composites based on animal fiber and their applications for a sustainable future. *Polymers*. 2023; *15*(3): 601. https://doi.org/10.3390/polym15030601

38. Koronis G, Silva A, Fontul M. Green composites: A review of adequate materials for automotive applications. *Composites Part B: Engineering*. 2013; 44(1): 120–7. https://doi.org/10.1016/j.compositesb.2012.07.004

39. Herrmann AS, Nickel J, Riedel U. Construction materials based upon biologically renewable resources – From components to finished parts. *Polymer Degradation and Stability*. 1998; 59: 251–61.

40. Alves C, Ferrão PMC, Silva AJ, Reis LG, Freitas M, Rodrigues LB. Eco-design of automotive components making use of natural jute fiber composites. *Journal of Cleaner Production*. 2011; 18: 313–27.

41. Jayaraman K. Manufacturing sisal–polypropylene composites with minimum fibre degradation. *Composites Science and Technology*. 2001; 63: 367–74.

42. Suddell B, Evans W. Natural fiber composites in automotive applications. In: Mohanty AK, Misra M, Drzal TL, editors. *Natural Fibers, Biopolymers, and Bio-Composites*. Boca Raton, Florida: CRC Press; 2005.

43. Mohanty AK, Misra M, Drzal TL, Selke SE, Harte BR, Hinrichsen G. Natural fibers, biopolymers, and bio-composites: An introduction. In: Mohanty AK, Misra M, Drzal TL, editors. *Natural Fibers, Biopolymers, and Bio-Composites*. Boca Raton: CRC Press-Taylor & Francis Group; 2005. pp. 1–36.

44. Knight M, Curliss D. *Encyclopedia of Physical Science and Technology* (3rd ed., Volume 3, pp. 455–68). Cambridge, MA, USA: Academic Press; 2003.

45. Kitane Y, Aref AJ. *Developments in Fiber-Reinforced Polymer (FRP) Composites for Civil Engineering*. Cambridge, UK: Woodhead Publishing; 15 May 2013.

46. Ogin SL, Brønsted P, Zangenberg, J. *Modeling Damage, Fatigue and Failure of Composite Materials*. Cambridge, UK: Woodhead Publishing; 2016.

47. Netravali AN, Chabba S. Composites get greener. *Materials Today*. 2003; 6: 22–26.

48. Marsh G. Next steps for automotive materials. *Materials Today*. 2003;6:36–43.

49. Luo S, Netravali AN. Mechanical and thermal properties of environment friendly "green" composites made from pineapple leaf fibers and poly(hydroxybutyrate-co-valerate) resin. *Polymer Composites*. 1999; 20: 367–78.

50. Khalid MY, Imran R, Arif ZU et al., Developments in chemical treatments, manufacturing techniques and potential applications of natural-fibers-based biodegradable composites. *Coatings*. 2021; 11(3): Article ID 293.

51. Lodha P, Netravali AN. Characterization of interfacial and mechanical properties of "green" composites with soy protein isolate and ramie fiber. *Journal of Materials Science*. 2002; 37: 3657–65.

52. Van de Velde K, Kiekens P. Biopolymers: Overview of several properties and consequences on their applications. *Polymer Testing*. 2002; 21:433–42.

53. Rodriguez-Gonzalez FJ, Ramsay BA, Favis BD. High performance LDPE/ thermoplastic starch blends: A sustainable alternative to pure polyethylene. *Polymer*. 2003; 44: 1517–26.

54. Scott G. Green polymers. *Polymer Degradation and Stability*. 2000; 68: 1–7.

55. Saad GR, Seliger H. Biodegradable copolymers based on bacterial poly((R)-3- hydroxyl-butyrate): Thermal and mechanical properties and biodegradation behaviour. *Polymer Degradation and Stability*. 2004; 83: 101–10.

56. Dufresne A, Vignon MR. Improvement of starch film performances using cellulose microfibrils. *Macromolecules*. 1998; 31:2693–6.

57. Lenz RW, Marchessault RH. Bacterial polyesters: Biosynthesis, biodegradable plastics and biotechnology. *Biomacromolecules*. 2005; 6: 1–8.

58. Godbole S, Gote S, Latkar M, Chakrabarti T. Preparation and characterization of biodegradable poly-3-hydroxybutyrate–starch blend films. *BioresourTechnol*. 2003; 86: 33–7.

59. Averous L. Biodegradable multiphase systems based on plasticized starch: A review. *Journal of Macromolecular Science*. 2004; 44: 231–74.

60. Yu L, Dean K, Li L. Polymer blends and composites from renewable resources. *Progress in Polymer Science*. 2006; 31: 576–602.

61. Satyanarayana KG, Arizaga GC, Wypych F. Biodegradable composites based on lignocellulosic fibers–An overview. *Progress in Polymer Science*. 2009; 34: 982–1021.

62. Khan MA, Idriss Ali KM, Hinrichsen G, Kopp C, Kropke S. Study on physical and mechanical properties of biopol-jute composite. *Polymer-Plastics Technology and Engineering*. 1999; 38: 99–112.

63. Mohanty AK, Khan MA, Sahoo S, Hinrichsen G. Effect of chemical modification on the performance of biodegradable jute yarn-Biopol composites. *Journal of Materials Science*. 2000; 35: 2589–95.

6 Biopolymers and Sustainable Biopolymers Based Composites
Fabrication Failure and Repairing

Jimmy Mehta and Prateek Mittal
Manav Rachna International Institute of Research
and Studies, Faridabad, India

Pallav Gupta
Amity School of Engineering and Technology,
Amity University, Noida, India

6.1 INTRODUCTION

Traditional petroleum-based plastics could potentially be replaced with biopolymers which are biodegradable and made from renewable resources like plants. New materials with improved characteristics have been created as a result of recent developments in the synthesis and processing of biopolymers. Biopolymers are made from carbon-neutral resources that are mostly made of bio-polymers such proteins and polysaccharides, which have greater biodegradability than synthetic polymers. A monomer, which is a building block made up of carbon (C), oxygen (O), hydrogen (H), and nitrogen (N), is generally repeated to create polymers. With regard to solid waste accretion, which is defined as growth or increase by the gradual accumulation of additional layers or matter, significant attention is being paid to the physio-chemical properties of polymers, such as stiffness, resilience, conductivity of heat and electricity, and resistance to degradation of biopolymers. Degradable biomaterials have undergone numerous attempts to be converted into monomers and/or carbon (C), oxygen (O), hydrogen (H), and/or nitrogen (N) without causing any negative environmental consequences.

Fabrication plays a crucial role in the context of biopolymers and composites, as it directly impacts the quality, performance, and sustainability of these materials. Biopolymers, derived from renewable resources, offer significant advantages over

DOI: 10.1201/9781003427735-6

traditional petroleum-based polymers in terms of environmental impact and resource depletion. However, successful fabrication is essential to harness their full potential. Fabrication processes such as compounding, extrusion, moulding, and 3D printing enable the shaping and formation of biopolymers into desired products, components, and structures. Proper fabrication ensures that biopolymers exhibit the desired mechanical, thermal, and chemical properties required for specific applications. In the case of composites, fabrication involves the integration of reinforcing materials, such as natural fibers or nano-particles, with biopolymer matrices to enhance their strength, stiffness, and durability. By carefully controlling fabrication parameters and techniques, it is possible to optimize material properties, achieve uniformity, and minimize defects. Furthermore, fabrication techniques contribute to the efficient use of resources, waste reduction, and the development of sustainable manufacturing practices. Thus, the importance of fabrication in the context of biopolymers and composites cannot be overstated, as it directly influences the overall performance, reliability, and environmental footprint of these materials. Bio-composites have gained significant traction in industrial applications, leading to the replacement of synthetic composites in various fields [1]. The automotive industry, in particular, has witnessed a substantial uptake of bio-composites, as evidenced by their widespread use [2–5]:

- Construction materials and building components (interior and exterior) [6–8]
- Furniture components and boards [9, 10]
- Sound absorbers for noise control [11]
- Mats, gardening articles, and storage cabinets [9]
- Packaging materials for electronics, foods, and other products [12–14]
- Biomedical and optical applications [15]
- Dentures, tissue engineering, medical implants, and 3d-printed joints [16–18]
- Marine application (limited) [19]

6.2 FABRICATION FAILURE IN BIOPOLYMERS AND COMPOSITES

Fabrication failure in biopolymers and composites can occur due to various factors and has significant implications for the performance and reliability of the materials. Several causes contribute to fabrication failure, including material-related factors such as degradation of biopolymers during processing or inadequate formulation of composite materials. Manufacturing process-related factors like improper curing or mixing can result in structural weaknesses and defects. Design-related factors, such as weak joints or inadequate support structures, can also lead to fabrication failure. The consequences of fabrication failure include reduced product performance, compromised durability, safety risks, and potential environmental impact. Fabrication failures can result in increased costs due to rework or replacement of components. Understanding and mitigating the causes of fabrication failure is critical to ensure the quality and integrity of biopolymers and composites throughout their lifecycle. This requires careful material selection, optimization of manufacturing processes, and proper design considerations to prevent failure and maximize the overall performance and reliability of these materials.

6.2.1 Causes of Fabrication Failure

6.2.1.1 Material-Related Factors

The handling of materials and the production process both have the potential to produce material flaws. Local hardness and other physical property deviations are brought on by the inclusion of material flaws and impurities. The physical characteristics of the material in the weld region will be impacted by the welding procedures used in fabrication. These issues are well known and can be prevented by using the right welding techniques and subsequent heat treating. Inspection methods can find faults in materials (refers to Figure 6.1). All of these are dependent on imperfect quality control. A number of enzymes are involved in the ester bond cleavage process, which breaks down biopolymers by hydrolysis; therefore, any faulty locations that are ignored are frequently the cause of corrosion [20].

The mechanism for the biodegradation of polymers occurs in phases, and each stage can be prevented. The sequence of the deterioration progression is shown in Figure 6.2.

With a semi-crystalline structure, polymers like bio-polyesters (PLA, PCL, PBS, PHB, cellulose, etc.) undergo a structural change around their glass transition temperatures, which alters the mobility of the macromolecular chains [21]. Due to the material's lower stiffness in the rubbery state (above Tg), the amorphous organization of the polymer chains promotes the degradation process. An increase in temperature could encourage the production of more crystallites, known as spherulites, in the glassy state

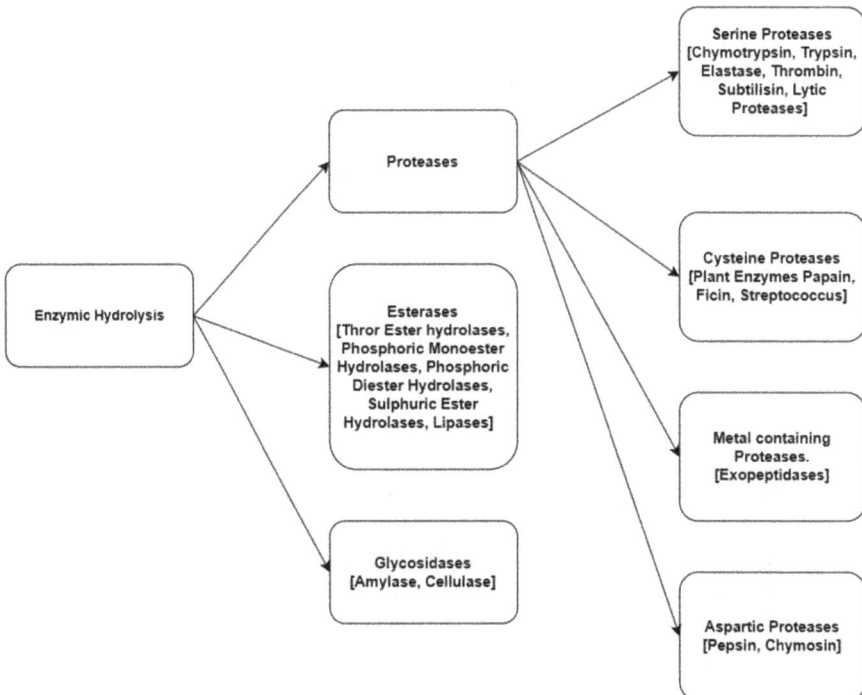

FIGURE 6.1 Types of enzymic hydrolysis.

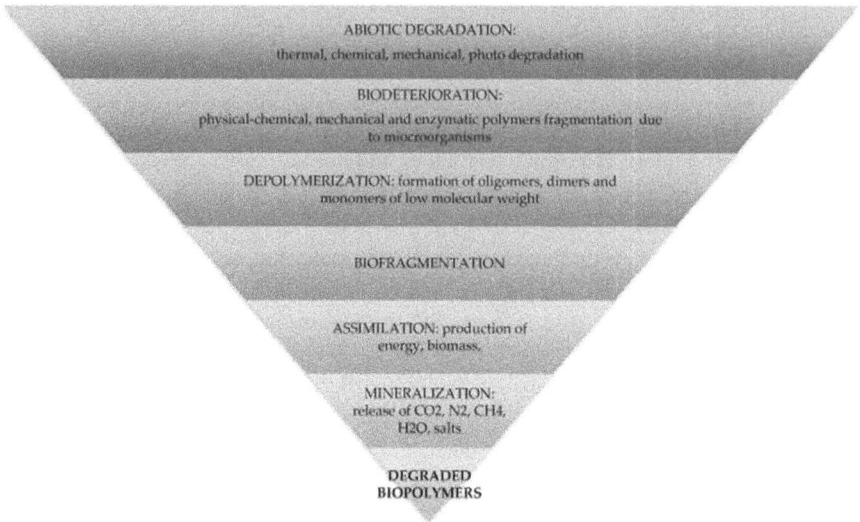

FIGURE 6.2   Degradation process sequence.

(sub Tg). There could be inter-spherulitic cracks that become more brittle, which would speed up the degradation process [22–25]. At the molecular level, mechanical damage caused by compression, tension, and shear forces can initiate or hasten the deterioration process [26].

6.2.2   MANUFACTURING PROCESS-RELATED FACTORS

During the manufacturing of structural composites, the occurrence of porosity or voids is a detrimental defect that significantly affects mechanical performance [27, 28]. Various researchers in the past have used different manufacturing practices to investigate the characteristics [29–53] of different materials. Voids are essentially air bubbles trapped within the composite matrix during fabrication [54], and their presence is influenced by factors such as curing pressure, resin system, and environmental conditions. Voids are commonly introduced into the material during the manufacturing process, and their formation and growth vary depending on the specific manufacturing technique used. In liquid composite molding, the focus is typically on void occurrence in the final process rather than voids that exist during the process itself [55, 56]. Void formation in liquid composite molding can be attributed to factors such as mechanical air entrapment, gas production from curing reactions, dissolved gases in the resin, and nucleation. The non-uniform permeability of fiber preforms, caused by the heterogeneous nature of the fibers, leads to local variations in resin velocity, exacerbating the capillary effect at the microscale [57, 58]. Additional considerations include:

- **Resin-rich zone**
  In the liquid composite molding process, resin-rich zones are a common issue that leads to undesired residual stress, deformation, and variations

between parts. These zones form during resin transfer in the molding process, where the dry fiber preform is compressed and a gap is created between the preform and mold surface. This compression causes the fibers to tighten around corners, resulting in resin-rich areas. Delamination and beam failure occur when these areas are present. To mitigate preform defects, vacuum assistance is necessary to reduce void content within the composites. Higher fiber contents can lead to wrinkling during preforming and mold closure.

- **Pocket of undispersed cross linker**
  Incomplete curing in composite structures can result in the formation of pockets of undispersed cross linker, which can be attributed to factors such as incomplete curing, uneven distribution of the curing agent, or premature curing. Before assessing the properties of composite structures, it is essential to evaluate the rheological and mechanical properties of the adhesive or epoxy used. Incomplete curing can significantly impact adhesion test results as it alters the characteristics of the adhesive. Under curing of thermoset composites ultimately leads to diminished performance properties in the final product.
- **Misaligned fibers**
  Fiber defects in polymer composites encompass various issues such as misalignment, wrinkles, waviness, folds, undulation, and breakage. Although there is currently no universally agreed-upon terminology for distinguishing between waves, wrinkles, and folds [59]. The presence of ply/fiber waviness or wrinkling caused by manufacturing processes, such as consolidation/curing, infiltration, and draping, often leads to reduced mechanical performance in composite parts. Fiber waviness refers to the deviation of plies or fibers from a straight alignment in unidirectional laminates. Out-of-plane fiber waviness, known as buckles or fiber buckling, occurs when a ply is subjected to compressive loading and experiences stability issues. Design-related factors, such as weak joints and inadequate support structures, significantly contribute to the failure of biopolymers. These factors weaken the structural integrity of biopolymer materials, leading to their inability to withstand applied forces and ultimately resulting in failure. It is essential to address these design considerations to enhance the performance and durability of biopolymer-based products.

6.3 CONSEQUENCES OF FABRICATION FAILURE

6.3.1 REDUCED PRODUCT PERFORMANCE AND DURABILITY

Fabrication failures in oil-based biopolymers can have severe consequences on product performance and durability. When fabrication processes are compromised, the resulting products may exhibit reduced mechanical strength, dimensional stability, and resistance to environmental factors. This can lead to premature wear, deformation, or even structural failure during use, compromising the overall functionality and lifespan of the biopolymer-based products. Inadequate fabrication can also negatively impact the material's chemical resistance, making it more susceptible to degradation when exposed to harsh environments or chemical agents. Moreover, fabrication failures may introduce defects such as voids, porosity, or weak bonding interfaces, further compromising the structural integrity of the biopolymer. These issues not only affect the performance and reliability of the products but also increase the risk of safety hazards and potential environmental damage. Therefore, ensuring proper fabrication techniques, quality control, and adherence to manufacturing standards are paramount to mitigate the consequences of fabrication failure and to maximize the performance and durability of oil-based biopolymers.

6.3.2 Safety Risks and Potential Environmental Impact

The production of sustainable biopolymers may involve the use of chemicals and processes that can pose hazards to workers and surrounding communities if not properly managed. For example, certain feed stocks or chemical additives used in biopolymer production may have toxicity concerns or require careful handling to avoid adverse health effects. Additionally, the introduction of new materials into existing waste streams may raise concerns about potential contamination and effects on recycling infrastructure. Regarding environmental impact, sustainable biopolymers can contribute to reducing reliance on fossil fuels and greenhouse gas emissions, especially when derived from renewable sources. However, their environmental footprint depends on various factors such as the raw materials used, energy consumption during production, and the disposal method chosen. For instance, if biopolymers are not effectively managed at the end of their lifecycle, they may still contribute to pollution and littering, potentially impacting ecosystems, wildlife, and aquatic environments. To minimize safety risks and potential environmental impact, it is crucial to implement responsible production practices, ensure proper handling and disposal methods, and promote effective waste management systems. Continued research and development efforts are necessary to address these concerns and optimize the sustainability and safety aspects of biopolymer materials throughout their lifecycle.

6.4 REPAIRING BIOPOLYMER-BASED COMPOSITES

6.4.1 Importance of Repair

Importance of repair in extending the lifecycle of biopolymer-based composites: Biopolymers offer a renewable alternative to traditional petroleum-based plastics and can be derived from a wide variety of feed-stocks, including agricultural products such as corn or soybeans and from alternative sources like algae or food waste. Biopolymer materials often exhibit inferior mechanical properties compared to petroleum-based polymers due to their inherent low stiffness. To overcome this limitation, reinforcing additives can be incorporated into biopolymers, derived from natural resources, to enhance their mechanical, physical, and electrical properties. Biopolymers also possess unique attributes such as self-healing capabilities, allowing them to restore their structural integrity in the event of failure as indicated in Figure 6.3. Bio-composites, for example, can withstand mechanical stress, maintain their size as designed in a culture environment, and are easily manipulated through suturing, gluing, or pin fixation. Utilizing sunflower oil as a self-healing agent, combined with bitumen sealant, has demonstrated improved resilience recovery and

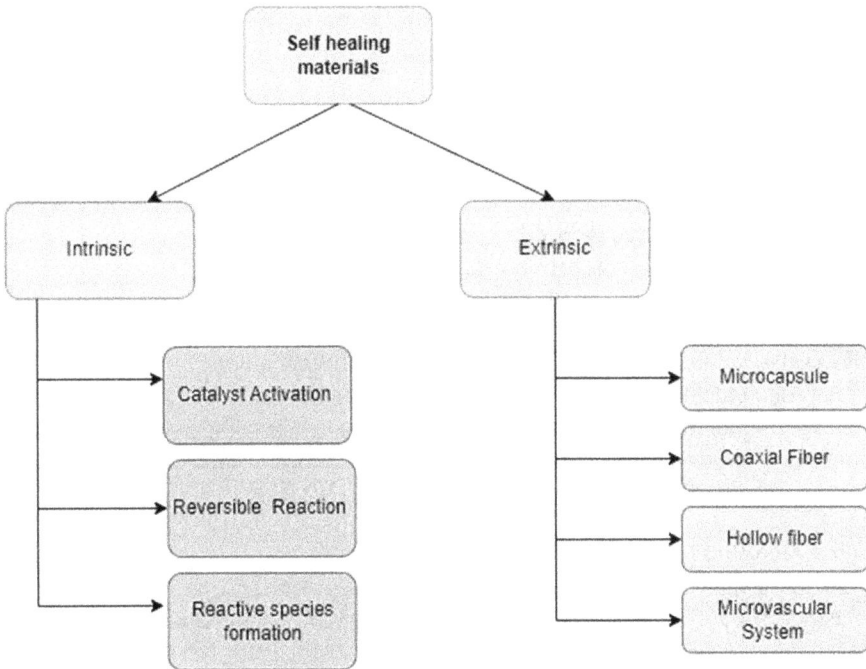

FIGURE 6.3   Self-healing materials.

fatigue life even at low temperatures. Self-healing can be achieved through auto-
nomic (without external intervention) or non-autonomic (with human intervention)
mechanisms. Controlling the release of healing agents, either through microcapsule
embedment or capillary-filled systems, is crucial when designing self-healing biopolymers or bio composites. Self-healing properties have also shown promise in
cartilage repair, addressing the limitations of conventional methods. Additionally,
self-healing enables internal crack repair under mechanical loading by releasing
chemicals that seal cracks and restore strength, ultimately retarding crack propagation. Factors such as dynamic loading, stimulus for chemical release, fiber reinforcement, monomer-carrying fibers, and chemical hardening methods need to be
carefully considered for effective self-repair (refers to Figure 6.3).

6.4.2   REPAIR TECHNIQUES AND METHODOLOGIES

6.4.2.1   Surface Preparation and Cleaning

Surface preparation and cleaning play a critical role in preventing biopolymer failure.
Proper surface preparation techniques, such as cleaning and roughening, enhance the
adhesion between biopolymer surfaces and coatings or adhesives, improving overall performance. Thorough cleaning removes contaminants, such as oils, dust, and
residues that can negatively impact the bonding and integrity of biopolymer surfaces.
Surface roughening techniques, such as mechanical or chemical treatments, create

microstructures that promote mechanical interlocking and increase surface area for better adhesion. By ensuring clean and properly prepared surfaces, biopolymer failure due to inadequate bonding, reduced adhesion strength, or premature delamination can be minimized, enhancing the reliability and longevity of biopolymer-based systems.

6.4.2.2 Adhesive Bonding and Composite Patching

Adhesive bonding and composite patching are essential methods used in biopolymer failure repair technology. Adhesive bonding involves using specially designed adhesives to create strong bonds between damaged biopolymer parts. Composite patching reinforces damaged areas with a composite patch consisting of reinforcing fibers embedded in a matrix material. Adhesive bonding provides seamless repairs without the need for additional materials and is flexible for different biopolymer components. Composite patching offers exceptional strength and stiffness for larger or heavily damaged structures. These techniques restore biopolymer components, extend their lifespan, and advancements in materials and technology continue to improve efficiency, sustainability, and cost-effectiveness.

6.4.2.3 Composite Material Reinforcement

Composite materials combine different materials to create a strong and versatile material. Biopolymers, derived from renewable sources, can be used as reinforcement in composites. While biopolymers offer benefits like biodegradability and reduced environmental impact, they have limitations in terms of strength and resistance to failure. To overcome these limitations, biopolymers are combined with a synthetic polymer matrix in composite materials. However, biopolymers can degrade under certain conditions and weaken the composite. Repair techniques involve using adhesives, resins, or additional reinforcement layers to restore strength and integrity. Careful assessment and selection of repair methods are crucial to maintain the mechanical properties of biopolymer-reinforced composites.

6.4.2.4 Additive Manufacturing and 3D Printing for Repairs

Additive manufacturing, or 3D printing, shows promising outcomes when employed for repairing bio-composite materials. Bio-composites consist of natural fibers in a biopolymer matrix. 3D printing repairs involve scanning the damaged area, creating a digital model, and depositing layers of compatible biopolymer material with reinforced natural fibers. This allows for customized, precise repairs and localized fixes, reducing waste. By incorporating natural fibers, the repaired component maintains compatibility and optimal mechanical performance. Additive manufacturing also enables the integration of functional elements. Challenges include ensuring proper adhesion and fiber orientation. Overall, additive manufacturing can improve the mechanical strength and durability of bio-composites, extending their lifespan.

6.4.3 Factors Influencing Repair Effectiveness

6.4.3.1 Compatibility Between Repair Materials and Biopolymers

The successful repair of biopolymers is influenced by various factors, with compatibility between the repair materials and the biopolymers being of paramount importance. Compatibility encompasses several aspects, including chemical composition, mechanical properties, and bonding characteristics. The repair materials should have similar chemical structures to the biopolymers to ensure effective bonding and prevent degradation or weakening. Mechanical properties, such as elasticity and stiffness, should be matched to maintain the structural integrity of the repaired biopolymers. Additionally, achieving strong adhesion between the repair materials and the biopolymers is crucial to ensure a durable and reliable repair. Careful consideration of these compatibility factors is essential for successful biopolymer repairs.

6.4.3.2 Adhesion Strength and Durability

Adhesion strength and durability are critical factors influencing the repair of biopolymers. Achieving strong adhesion between the repair materials and the biopolymers is essential to ensure a reliable and long-lasting repair. Factors that affect adhesion include surface preparation, compatibility between the repair materials and the biopolymers, and the bonding mechanism employed. Proper surface treatment, such as cleaning or roughening, improves adhesion by enhancing the contact area between the materials. Compatibility in terms of chemical composition and intermolecular interactions promotes strong bonding. Furthermore, the selection of appropriate bonding mechanisms, such as chemical adhesives or interlocking structures, can significantly impact the adhesion strength and durability of biopolymer repairs.

6.4.3.3 Post-repair Testing and Evaluation

Post-repair testing and evaluation play a crucial role in assessing the success and effectiveness of biopolymer repairs. Several factors influence this process. First, selecting appropriate testing methods is important to evaluate the mechanical properties, such as strength, stiffness, and impact resistance, of the repaired biopolymers. Additionally, the choice of evaluation criteria, including industry standards or specific performance requirements, helps determine whether the repaired biopolymers meet the desired specifications. Moreover, considering the long-term durability and stability of the repaired structure through accelerated aging tests or exposure to relevant environmental conditions ensures the reliability and performance of the repair. Proper post-repair testing and evaluation enable informed decisions and ensure the quality of biopolymer repairs.

6.5 STRATEGIES FOR PREVENTING FABRICATION FAILURE

6.5.1 MATERIAL SELECTION AND CHARACTERIZATION

6.5.1.1 Understanding Biopolymer Properties and Limitations

Biopolymers are natural polymers derived from renewable sources such as plants, animals, and microorganisms. Understanding their properties and limitations is crucial for their successful application in various fields. Biopolymers exhibit diverse characteristics, including biodegradability, biocompatibility, and tunable mechanical properties. However, they often possess lower thermal stability and mechanical strength compared to synthetic polymers. Additionally, their high cost of production, limited processing techniques, and susceptibility to microbial degradation pose challenges. In-depth knowledge of biopolymer properties and limitations enable researchers to optimize their formulation, processing methods, and end-use applications, promoting sustainable and environmentally friendly solutions.

6.5.1.2   Evaluating Sustainable Additives and Reinforcements

Evaluating sustainable additives and reinforcements is crucial for preventing fabrication failure and enhancing the material performance of biopolymers. Sustainable additives, such as natural fillers, plasticizers, and compatibilizers, can improve the mechanical properties, process ability, and stability of biopolymer materials. These additives can enhance the interfacial adhesion between the biopolymer matrix and reinforcements, such as natural fibers, nanoparticles, or biodegradable microspheres. Reinforcements offer improved strength, stiffness, and impact resistance to biopolymers. However, it is essential to consider the compatibility between the reinforcements and biopolymers to ensure proper dispersion and effective load transfer. Evaluating the performance of sustainable additives and reinforcements can prevent fabrication failures, such as poor bonding, delamination, or structural instability, while promoting the use of environmentally friendly materials in various applications.

6.5.2   ESTABLISHING PROPER PROCESSING PARAMETERS

It is crucial for quality control in biopolymer production. Relevant process parameters include factors such as temperature, pressure, residence time, and cooling rates. Controlling these parameters ensures the desired material properties, such as molecular weight, crystallinity, and mechanical strength. Additionally, proper processing parameters prevent issues like thermal degradation, inadequate mixing, or poor dispersion of additives. Techniques like melt blending, extrusion, and injection molding require precise control to achieve consistent and high-quality biopolymer products. By establishing and monitoring the appropriate processing parameters, manufacturers can optimize the performance and functionality of biopolymers, ensuring their suitability for various applications while maintaining consistency and meeting quality standards.

6.5.3   MONITORING AND CONTROLLING FABRICATION CONDITIONS

This is one of the most crucial fabrication conditions for ensuring enhanced structural integrity and preventing biopolymer failure in the design process and quality control. Fabrication conditions refer to the parameters and variables involved in development of biopolymer structure, such as temperature, humidity, curing time, and processing techniques. Monitoring the fabrication conditions allows for real-time assessment of the process and ensures that the biopolymers are being produced under optimal conditions. Sensors and monitoring systems are employed to track parameters like temperature and humidity levels throughout the fabrication process. Controlling the fabrication conditions involves adjusting and maintaining the parameters within desired ranges. For example, maintaining a specific curing temperature and duration ensures proper crosslinking and structural stability of the biopolymer. Controlling the processing techniques, such as extrusion or injection molding, helps in achieving uniform material distribution and minimizing defects. By monitoring and controlling the fabrication conditions, it is possible to optimize the manufacturing process, improve material properties, and enhance the structural integrity of the biopolymer design. This leads to reduced chances of failure and ensures that the final product meets the required performance standards.

6.6 DESIGN CONSIDERATIONS FOR ENHANCED STRUCTURAL INTEGRITY

6.6.1 INCORPORATING EFFICIENT JOINT DESIGNS AND REINFORCEMENT STRATEGIES

Efficient joint designs involve cautious consideration of factors like stress concentration, material compatibility, and joint geometry. High-stress concentration areas should be minimized or redistributed to prevent failure. Additionally, choosing appropriate joining techniques, such as adhesive bonding or mechanical fastening, based on the properties of the biopolymer and the intended application, is crucial.

Reinforcement strategies play a vital role in enhancing the overall strength and stability of biopolymer structures. This is achieved by incorporating highlighting elements, such as fibers, particles, or fillers, into the biopolymer matrix. These reinforcements improve mechanical properties such as stiffness, strength, and toughness. The reinforcement can be uniformly dispersed throughout the material or strategically placed in regions of higher stress to improve load distribution.

Besides, incorporating reinforcement strategies like ribbing, honeycomb structures, or lattice frameworks can effectively distribute loads and enhance structural integrity. These designs provide additional support and prevent localized stress concentration, thereby reducing the risk of failure. By incorporating efficient joint designs and reinforcement strategies, biopolymer structures can achieve enhanced structural integrity, improved load-bearing capacity, and increased resistance to failure. These considerations are vital in various industries where biopolymers are used, including automotive, aerospace, and biomedical application.

6.6.2 IMPLEMENTING PROPER SUPPORT STRUCTURES AND LOAD DISTRIBUTION

Self-healing biopolymers restore their structural integrity and withstand applied mechanical load so as to keep their designed structure intact and does not shrink. The process involves careful consideration of material selection, structural design, and simulation/testing. Appropriate biopolymer material must be chosen based on mechanical properties together with tensile strength, flexibility, and environmental resistance. This guarantees the biopolymer can withstand anticipated loads and circumstances. Also, structural design should integrate support structures that distribute loads evenly. Reinforcing elements like ribs or struts can be added to strengthen weak areas and prevent stress concentrations. The shape and geometry of the structure should also be optimized to minimize stress concentrations and promote load distribution. Simulation and testing play a vital role in validating the design. Computer simulations using finite element analysis can assess the performance of the structure under various loading scenarios. Physical testing provides real-world validation of the design's integrity and identifies any necessary modifications. By implementing proper support structures and load distribution techniques, biopolymer designs can achieve enhanced structural integrity, ensuring they can withstand expected loads and prevent failure.

6.7 FUTURE DIRECTIONS AND EMERGING TECHNOLOGIES

6.7.1 ADVANCEMENTS IN BIOPOLYMER RESEARCH AND DEVELOPMENT

The industrial interest in biopolymers has been growing steadily due to the demand for sustainable and cost-effective materials. Bio-based polymers are increasingly being used in various applications, including food packaging, medicine, and agriculture, as they offer advantages such as biodegradability and renewable resources. Traditional synthetic polymers used in these applications are non-degradable and contribute to environmental pollution. Biodegradable polymers, derived from renewable sources such as cellulose, starch, and proteins, have the potential to replace petroleum-based materials and address environmental concerns. Researchers are actively developing and testing various types of biodegradable polymers to evaluate their efficacy, safety, and environmental impact. Additionally, bio-based polymers are being explored as replacements for existing polymers through bacterial fermentation processes using renewable resources. The future focus is on making plastics more biodegradable while maintaining their strength and durability, which would revolutionize the plastics industry and promote a sustainable lifestyle. The market for biodegradable plastics is expected to grow significantly in the coming years.

6.7.2 NOVEL REPAIR TECHNIQUES AND MATERIALS

Fiber-reinforced polymers are increasingly utilized in high-stress applications such as aviation, vehicle construction, shipbuilding, and mechanical engineering due to their exceptional specific properties. Consequently, great emphasis is placed on repair technologies specifically designed for composites, enabling the preservation and reproducible restoration of mechanical properties. Therefore, an innovative repair technology is being examined, utilizing form-closed and adhesive bond interlock elements [60]. This interlock technology is customized for one-sided automated preparation of repair zones and subsequent repair of composite structures. As a result, it enables the achievement of minimal repair zones, even for thick laminates, while preserving the surface quality.

6.7.3 INTEGRATION OF SMART MATERIALS AND SELF-HEALING CAPABILITIES

Self-healing materials have a long history [61], with concrete constructions being the first to exhibit self-healing characteristics over 2000 years ago. Since then, self-healing has been seen in a variety of materials, including elastomers [62], shape memory polymers [63], thermoplastic polymers [64], thermoset polymers [65], polymer composites, and nano-composites [66]. The development of multifunctional materials that can regain their mechanical strength, electrical conductivity, and corrosion resistance is the main goal of research on synthetic self-healing materials [67]. The spectrum of applications for these materials may increase with the development of self-healing materials.

Depending on the chemistries at play, the mechanisms of self-healing can be classified as extrinsic or intrinsic [68]. While intrinsic self-healing can take place without external agents, extrinsic self-healing requires them in the form of capsules or vesicles. Both times, material damage triggers the self-healing procedure. While intrinsic self-healing relies on non-covalent chemistries like – stacking, hydrogen bonding, and host-guest interactions, extrinsic self-healing uses microcapsules carrying healing chemicals.

In comparison to non-healing materials, self-healing polymers and nanocomposites have better physical properties [69], supramolecular forces, and structural durability. They have been used in a variety of industries, including aircraft, coatings, electronics, and energy. Self-healing nano-composites have been developed for the aerospace industry employing nanoparticles and micro- or nano-capsules as healing agents [70] Self-recovering aerospace nanocomposites have been made possible by the use of polymers and nano-carbon nanofillers. Innovative design strategies are required for future developments in self-healing nanocomposites for aeronautical constructions in order to increase healing effectiveness [71–73].

6.8 CONCLUSION

Fabrication and repairing processes play a crucial role in the development and utilization of biopolymers and composites. Proper fabrication methods ensure the desired properties and performance of biopolymers by controlling factors such as filler dispersion, interfacial interactions, and processing conditions. Additionally, repairing mechanisms, such as self-healing capabilities, contribute to the longevity and durability of biopolymer-based materials by restoring structural integrity and mitigating damage. Both fabrication and repairing techniques are essential for optimizing the mechanical, physical, and ecological properties of biopolymers and composites, thereby enabling their effective use in various applications and promoting a sustainable and efficient material economy. Biopolymer composites have emerged as a promising area of development for sustainable economies, offering sociological, ecological, and property improvements. By reducing dependence on petroleum-based materials, bio-composites significantly decrease greenhouse gas emissions. However, the transition from traditional materials to renewable alternatives is a complex process. This study emphasizes the importance of proper biopolymer selection and fabrication methods to enhance crucial properties. Achieving homogeneous filler dispersion within the biopolymer matrix improves mechanical performance. Effective filler selection, treatment, and interfacial engineering further enhance mechanical properties. Strong bonding interactions between modified filler surfaces and biopolymers contribute to high modulus and other mechanical attributes. Optimal interfacial adhesion, influenced by factors like filler aspect ratio and volume fraction, enhances composite strength. It is essential for the research community to prioritize the development of biopolymer composites suitable for mechanical applications while ensuring minimal impact on health and the environment. Governments worldwide are increasing funding and research efforts to advance the efficiency of biopolymer composites, positioning them as future materials for various applications. Potential impact of advancements in the field on sustainability and industry applications. The interest in biopolymers has grown due to their abundant availability, compostability, low cost, and ecological advantages over synthetic materials. Factors driving the adoption of compostable biopolymers include restrictions on conventional plastics, bio-based product demand, and zero-waste initiatives. Nature offers a wide range of biopolymers with varying properties and molecular structures, which can be controlled through processing techniques and parameters. The performance of biopolymers is influenced by chemical composition, physical properties, modification techniques, additives, structural defects, and environmental conditions. Technological advancements have facilitated the scaling up of biopolymer applications, bringing their performance closer to that of petroleum-based synthetic polymers. However, further research is needed to explore processing techniques and develop new methods for biopolymer utilization.

REFERENCES

1. Campilho RDSG (2017). Recent innovations in biocomposite products. In *Biocomposites for high-performance applications* (pp. 275–306). Woodhead Publishing. https://www.sciencedirect.com/book/9780081007938/biocomposites-for-high-performance-applications
2. Guo Y, & Deng Y (2020). Recycling of flax fiber towards developing biocomposites for automotive application from a life cycle assessment perspective. In *Reference module in materials science and materials engineering*. https://doi.org/10.1016/B978-0-12-803581-8.11496-1
3. Akampumuza O, Wambua PM, Ahmed A, Li W, & Qin XH (2017). Review of the applications of biocomposites in the automotive industry. *Polymer Composites*, *38*(11), 2553–2569.
4. Khalfallah M, Abbès B, Abbès F, Guo YQ, Marcel V, Duval A, ... & Rousseau F (2014). Innovative flax tapes reinforced Acrodur biocomposites: A new alternative for automotive applications. *Materials & Design*, *64*, 116–126.
5. Roy P, Tadele D, Defersha F, Misra M, & Mohanty AK (2019). Environmental and economic prospects of biomaterials in the automotive industry. *Clean Technologies and Environmental Policy*, *21*, 1535–1548.
6. Yan L, & Chouw N (2013). Experimental study of flax FRP tube encased coir fibre reinforced concrete composite column. *Construction and Building Materials*, *40*, 1118–1127.
7. Guadagnuolo M, & Faella G (2020). Simplified design of masonry ring-beams reinforced by flax fibers for existing buildings retrofitting. *Buildings*, *10*(1), 12.
8. Korjenic A, Zach J, & Hroudová J (2016). The use of insulating materials based on natural fibers in combination with plant facades in building constructions. *Energy and Buildings*, *116*, 45–58.
9. Bharath KN, & Basavarajappa S (2016). Applications of biocomposite materials based on natural fibers from renewable resources: A review. *Science and Engineering of Composite Materials*, *23*(2), 123–133.
10. Grozdanov A, Jordanov I, Errico ME, Gentile G, & Avella M (2015). Biocomposites based on natural fibers and polymer matrix—From theory to industrial products. *Green Biorenewable Biocomposites From Knowledge to Industrial Applications*, 323–344. https://doi.org/10.1201/b18092

11. Yilmaz ND, & Powell NB (2015). Biocomposite structures as sound absorber materials. In VijayKumar Thakur, and MichaelR Kessler (Eds.), *Green biorenewable biocomposites: From knowledge to industrial applications*. Boca Rayton, FL: Routledge.

12. Sirviö JA, Kolehmainen A, Liimatainen H, Niinimäki J, & Hormi OE (2014). Biocomposite cellulose-alginate films: Promising packaging materials. *Food chemistry*, *151*, 343–351.

13. Marra A, Silvestre C, Duraccio D, & Cimmino S (2016). Polylactic acid/zinc oxide biocomposite films for food packaging application. *International Journal of Biological Macromolecules*, *88*, 254–262.

14. Narayanan M, Loganathan S, Valapa RB, Thomas S, & Varghese TO (2017). UV protective poly (lactic acid)/rosin films for sustainable packaging. *International Journal of Biological Macromolecules*, *99*, 37–45.

15. Annamalai PK, & Depan D (2015). Nano-cellulose reinforced chitosan nanocomposites for packaging and biomedical applications. In VijayKumar Thakur, and MichaelR Kessler (Eds.), *Green biorenewable biocomposites: from knowledge to industrial applications* (pp. 489–506). Routledge.

16. Fouly A, Ibrahim AMM, Sherif ESM, FathEl-Bab AM, & Badran AH (2021). Effect of low hydroxyapatite loading fraction on the mechanical and tribological characteristics of poly (methyl methacrylate) nanocomposites for dentures. *Polymers*, *13*(6), 857.

17. Fouly A, Alnaser IA, Assaifan AK, & Abdo HS (2022). Evaluating the performance of 3D-printed PLA reinforced with date pit particles for its suitability as an acetabular liner in artificial hip joints. *Polymers*, *14*(16), 3321.

18. Ilyas RA, Zuhri MYM, Norrrahim MNF, Misenan MSM, Jenol MA, Samsudin SA, … & Omran AAB (2022). Natural fiber-reinforced polycaprolactone green and hybrid biocomposites for various advanced applications. *Polymers*, *14*(1), 182.

19. Reddy TRK, Kim HJ, & Park JW (2016). Renewable biocomposite properties and their applications. In *Composites from renewable and sustainable materials* (p. 177). https://doi.org/10.5772/65475

20. Wong W (2010). Asset integrity: learning about the cause and symptoms of age and decay and the need for maintenance to avoid catastrophic failures. *Risk management of safety and dependability* (pp. 188–225). https://doi.org/10.1533/9781845699383.188

21. Jakubowicz I, Yarahmadi N, & Petersen H (2006). Evaluation of the rate of abiotic degradation of biodegradable polyethylene in various environments. *Polymer Degradation and Stability*, *91*(7), 1556–1562.

22. Siracusa V, Dalla Rosa M, & Iordanskii AL (2017). Performance of poly (lactic acid) surface modified films for food packaging application. *Materials*, *10*(8), 850.

23. Siracusa V (2019). Microbial degradation of synthetic biopolymers waste. *Polymers*, *11*(6), 1066.

24. Iovino R, Zullo R, Rao MA, Cassar L, & Gianfreda L (2008). Biodegradation of poly (lactic acid)/starch/coir biocomposites under controlled composting conditions. *Polymer Degradation and Stability*, *93*(1), 147–157.

25. El-Hadi A, Schnabel R, Straube E, Müller G, & Henning S (2002). Correlation between degree of crystallinity, morphology, glass temperature, mechanical properties and biodegradation of poly (3-hydroxyalkanoate) PHAs and their blends. *Polymer Testing*, *21*(6), 665–674.

26. Briassoulis D (2005). The effects of tensile stress and the agrochemical Vapam on the ageing of low density polyethylene (LDPE) agricultural films. Part I. Mechanical behaviour. *Polymer Degradation and Stability*, *88*(3), 489–503.

27. Mehdikhani M, Gorbatikh L, Verpoest I, & Lomov SV (2019). Voids in fiber-reinforced polymer composites: A review on their formation, characteristics, and effects on mechanical performance. *Journal of Composite Materials*, *53*(12), 1579–1669.

28. Ornaghi Jr HL, Neves RM, Monticeli FM, & Almeida Jr JHS (2020). Viscoelastic characteristics of carbon fiber-reinforced epoxy filament wound laminates. *Composites Communications*, *21*, 100418.

29. Sharma A, Kumar V, Babbar A, Dhawan V, Kotecha K, & Prakash C (2021 Jan). Experimental investigation and optimization of electric discharge machining process parameters using grey-fuzzy-based hybrid techniques. *Materials*, 14(19), 5820.

30. Prakash C, Kumar V, Mistri A, Uppal AS, Babbar A, Pathri BP, Mago J, Sharma A, Singh S, Wu LY, & Zheng HY (2021 Jan). Investigation of functionally graded adherents on failure of socket joint of FRP composite tubes. *Materials*, *14*(21), 6365.

31. Babbar A, Jain V, Gupta D, & Prakash C (2021 Jul). Experimental investigation and parametric optimization of neurosurgical bone grinding under bio-mimic environment. *Surface Review and Letters*, *28*, 2141005.

32. Babbar A, Jain V, Gupta D, Agrawal D, Prakash C, Singh S, Wu LY, Zheng HY, Królczyk G, & Bogdan-Chudy M (2021 Feb 24). Experimental analysis of wear and multi-shape burr loading during neurosurgical bone grinding. *Journal of Materials Research and Technology*, *12*, 15–28.

33. Babbar A, Jain V, Gupta D, & Agrawal D (2021 Feb 16). Histological evaluation of thermal damage to osteocytes: A comparative study of conventional and ultrasonic-assisted bone grinding. *Medical Engineering & Physics*, *90*, 1–8.

34. Babbar A, Jain V, Gupta D, & Agrawal D (2021 Feb 16). Finite element simulation and integration of CEM43°C and Arrhenius models for ultrasonic-assisted skull bone grinding: A thermal dose model. *Medical Engineering & Physics*, *90*, 9–22.

35. Babbar A, Prakash C, Singh S, Gupta MK, Mia M, and Pruncu CI (2020 Jul 1). Application of hybrid nature-inspired algorithm: Single and bi-objective constrained optimization of magnetic abrasive finishing process parameters. *Journal of Materials Research and Technology*, *9*(4), 7961–7974.

36. Baraiya R, Babbar A, Jain V, and Gupta D (2020 Feb 1). In-situ simultaneous surface finishing using abrasive flow machining via novel fixture. *Journal of Manufacturing Processes*, *50*, 266–278.

37. Singh S, Prakash C, Pramanik A, Basak A, Shabadi R, Królczyk G, Bogdan-Chudy M, & Babbar A (2020 Jan). Magneto-rheological fluid assisted abrasive nanofinishing of β-phase Ti-Nb-Ta-Zr alloy: Parametric appraisal and corrosion analysis. *Materials*, *13*(22), 5156.

38. Babbar A, Jain V, & Gupta D (2020 Jun 23). Preliminary investigations of rotary ultrasonic neurosurgical bone grinding using Grey-Taguchi optimization methodology. *Grey Systems: Theory and Application*, *40*, 479–493.

39. Babbar A, Jain V, & Gupta D (2020 Mar 17). In vivo evaluation of machining forces, torque, and bone quality during skull bone grinding. *Proceedings of the Institution of Mechanical Engineers, Part H: Journal of Engineering in Medicine*, *234*, 626–638.

40. Babbar A, Jain V, & Gupta D (2019 Oct 1). Thermogenesis mitigation using ultrasonic actuation during bone grinding: a hybrid approach using CEM43°C and Arrhenius model. *Journal of the Brazilian Society of Mechanical Sciences and Engineering*, 41(10), 401. (SCI IF = 1.755)

41. Babbar A, Sharma A, Jain V, & Jain AK (2019 Apr 24). Rotary ultrasonic milling of C/SiC composites fabricated using chemical vapor infiltration and needling technique. *Materials Research Express*. (SCI IF = 1.929) ISSN: 2053–1591, 08-05-2019. https://doi.org/10.1088/2053-1591/ab1bf7

42. Singh G, Babbar A, Jain V, & Gupta D (2021 May 1). Comparative statement for diametric delamination in drilling of cortical bone with conventional and ultrasonic assisted drilling techniques. *Journal of Orthopaedics*, 25, 53–58.

43. Babbar A, Sharma A, & Singh P (2022 Jan 1). Multi-objective optimization of magnetic abrasive finishing using grey relational analysis. *Materials Today: Proceedings*, 50, 570–575.

44. Sharma A, Kalsia M, Uppal AS, Babbar A, & Dhawan V (2022 Jan 1). Machining of hard and brittle materials: A comprehensive review. *Materials Today: Proceedings*, 50, 1048–1052.

45. Babbar A, Sharma A, Bansal S, Mago J, & Toor V (2020 Jan 1). Potential applications of three-dimensional printing for anatomical simulations and surgical planning. *Materials Today: Proceedings*, 33, 1558–1561.

46. Babbar A, Jain V, & Gupta D (2020 Jan 1). Thermo-mechanical aspects and temperature measurement techniques of bone grinding. *Materials Today: Proceedings*, 33, 1458–1462.

47. Sharma A, Babbar A, Jain V, Gupta D Enhancement of surface roughness for brittle material during rotary ultrasonic machining. In MATEC Web of Conferences 2018 (Vol. 249, p. 01006). EDP Sciences.

48. Babbar A, Singh P, & Farwaha HS (2017 Aug). Regression model and optimization of magnetic abrasive finishing of flat brass plate. *Indian J Sci Technol*, 10, 1–7.

49. Babbar A, Singh P, & Farwaha HS (2017). Parametric Study of magnetic abrasive finishing of UNS c26000 flat brass Plate. *Int J Adv Mechatronics Robot*, 9, 83–89.

50. Babbar A, Sharma A, & Chugh M. (2020). Application of flexible sintered magnetic abrasive brush for finishing of brass plate. *Optimization in Engineering Research* 1(1), 36–47.

51. Babbar A, Jain V, Gupta D, Prakash C, & Agrawal D (2022). Potential application of CEM43° C and Arrhenius model in neurosurgical bone grinding. In Chander Prakash, Sunpreet Singh, Aminesh Basak, J. Paulo Davim (Eds.), *Numerical modelling and optimization in advanced manufacturing processes* (pp. 145–158). Cham: Springer.

52. Babbar A, Jain V, & Gupta D (2019). Neurosurgical bone grinding. In Prakash C et al. (Eds.), *Biomanufacturing*. Cham (Scopus indexed): Springer.

53. Sharma A, Grover V, Babbar A, & Rani R (2020 Oct 30). A trending nonconventional hybrid finishing/machining process. In Chander Prakash, Sunpreet Singh, Aminesh Basak, J. Paulo Davim (Eds.), *Non-conventional hybrid machining processes* (pp. 79–93).

54. Suriani MJ, Ali A, Khalina A, Sapuan SM, & Abdullah S (2012). Detection of defects in kenaf/epoxy using infrared thermal imaging technique. *Procedia Chemistry, 4*, 172–178.

55. Xueshu L., & Fei C. (2016). A review of void formation and its effects on the mechanical performance of carbon fiber reinforced plastic. *Engineering Transactions, 64*(1), 33–51.

56. Lundstrom TS, Gebart BR, & Lundemo CY (1993). Void formation in RTM. *Journal of Reinforced Plastics and Composites, 12*(12), 1339–1349.

57. Afendi MD, Banks WM, & Kirkwood D (2005). Bubble free resin for infusion process. *Composites Part A: Applied Science and Manufacturing, 36*(6), 739–746.

58. Kang MK, Lee WI, & Hahn HT (2000). Formation of microvoids during resin-transfer molding process. *Composites Science and Technology, 60*(12–13), 2427–2434.

59. Wemyss AM, Ellingford C, Morishita Y, Bowen C, & Wan C (2021). Dynamic polymer networks: A new avenue towards sustainable and advanced soft machines. *Angewandte Chemie International Edition, 60*(25), 13725–13736.

60. Li QH, Yin X, Huang BT, Luo AM, Lyu Y, Sun CJ, & Xu SL (2021). Shear interfacial fracture of strain-hardening fiber-reinforced cementitious composites and concrete: A novel approach. *Engineering Fracture Mechanics, 253*, 107849.

61. Tan YJ, Susanto GJ, Anwar Ali HP, & Tee BC (2021). Progress and Roadmap for Intelligent Self-Healing Materials in Autonomous Robotics. *Advanced Materials*, *33*(19), 2002800.

62. Utrera-Barrios S, Verdejo R, López-Manchado MÁ, & Santana MH (2022). The final frontier of sustainable materials: Current developments in self-healing elastomers. *International Journal of Molecular Sciences*, *23*(9), 4757.

63. Liu X, Zhang E, Liu J, Qin J, Wu M, Yang C, & Liang L (2023). Self-healing, reprocessable, degradable, thermadapt shape memory multifunctional polymers based on dynamic imine bonds and their application in nondestructively recyclable carbon fiber composites. *Chemical Engineering Journal*, *454*, 139992.

64. Chalapathi KV, Prabhakar MN, Lee DW, & Song JI (2023). Development of thermoplastic self-healing panels by 3D printing technology and study extrinsic healing system under low-velocity impact analysis. *Polymer Testing*, *119*, 107923.

65. Zhang F, Zhang L, Yaseen M, & Huang K (2021). A review on the self-healing ability of epoxy polymers. *Journal of Applied Polymer Science*, *138*(16), 50260.

66. Kim YN, Jeong H, Ryu S, & Jung YC (2023). Self-healing polymers and composites for additive manufacturing: Materials, properties, and applications. *Nanotechnology-Based Additive Manufacturing: Product Design, Properties and Applications*, *1*, 219–248.

67. Fatahi A, Mohammadi A, & Sarfjoo MR (2023). Polymers for smart coatings. In *Specialty polymers* (pp. 233–246). Boca Rayton, FL: CRC Press.

68. Pezzin, SH (2023). Mechanism of extrinsic and intrinsic self-healing in polymer systems. In *Multifunctional epoxy resins: Self-healing, thermally and electrically conductive resins* (pp. 107–138). Singapore: Springer Nature Singapore.

69. Klingler WW, Bifulco A, Polisi C, Huang Z, & Gaan S (2023). Recyclable inherently flame-retardant thermosets: Chemistry, properties and applications. *Composites Part B: Engineering*, *258*, 110667.

70. Parameswaran B, & Singha NK (2022). 15 Self-healing composites – Capsule and vascular based extrinsic self-healing systems. In Sri Bandyopadhyay, & Raghavendra Gujjala (Eds.), *Toughened composites: micro and macro systems* (p. 203). Boca Rayton, FL: Routledge.

71. Wemyss AM, Ellingford C, Morishita Y, Bowen C, & Wan C (2021). Dynamic polymer networks: A new avenue towards sustainable and advanced soft machines. *Angewandte Chemie International Edition*, *60*(25), 13725–13736.

72. Paolillo S, Bose RK, Santana MH, & Grande AM (2021). Intrinsic self-healing epoxies in polymer matrix composites (PMCs) for aerospace applications. *Polymers*, *13*(2), 201.

73. Ekeocha J, Ellingford C, Pan M, Wemyss AM, Bowen C, & Wan C (2021). Challenges and opportunities of self-healing polymers and devices for extreme and hostile environments. *Advanced Materials*, *33*(33), 2008052.

7 Biomimicry-Inspired Design of Sustainable Composite Materials

Vidyapati Kumar
Indian Institute of Technology, Kharagpur, India

Ankita Mistri
Indian Institute of Technology, Dhanbad, India

Atul Babbar and Vikas Dhawan
Shree Guru Gobind Singh Tricentenary University,
Gurugram, India

Raman Kumar
Guru Nanak Dev Engineering College, Ludhiana, India

Lavish Kumar Singh
Sharda University, Greater Noida, India

Ankit Sharma
Chitkara University, Rajpura, India

7.1 INTRODUCTION

In the captivating journey of scientific exploration, nature has consistently remained a profound source of inspiration, guiding humanity's quest for ingenious solutions. Over centuries, mankind has diligently observed and harnessed the peculiar adaptations and thriving strategies that natural organisms and ecosystems have evolved to survive. By delving into the intricate geometries, mechanisms, and behaviours of the living world, profound insights have been gained, paving the way for transformative innovations. Manufacturing operations have always been a key-point for materials fabrication and processing [1–29].

The ancient wisdom of civilizations such as the Egyptians, who crafted temples inspired by the graceful lotus plants, and the modern brilliance of visionaries like Gaudi, who engineered structures influenced by natural topologies such as catenary arches and optimized forms [30], exemplify the timeless allure of mimicking nature's finest creations. This pursuit culminated in the advent of a groundbreaking term in 1982 that resonated across the scientific community—biomimicry. It encapsulated the compelling approach of emulating nature's genius to address human needs. As the years passed, the concept expanded and gained clarity, defining biomimicry as a method that endeavours to develop techniques and technologies that imitate nature's genius or derive inspiration from specific features observed within it [31, 32]. This fascinating exploration of biomimicry looks at three levels of studying and emulating nature – individual organisms, behaviours, and entire ecosystems. Each level provides unique perspectives into the art of learning from the living world. We will closely examine five key characteristics: form, material composition, hierarchical structure, functions, and operating principles. These aspects serve as engaging gateways to unravel the mysteries of nature, unlocking its secrets to enable a harmonious coexistence with our planet. Engineered and biological systems differ considerably in many aspects, making a direct replica of nature impossible. As shown in Table 7.1, the differences encompass not just the building blocks but also diverse attributes. From multi-level hierarchies to self-healing capabilities, natural materials have adapted specific shapes and architectures tailored to particular roles rather than being designed for pre-defined applications. This highlights the challenges in directly

TABLE 7.1
Contrasting Key Attributes of Materials from Nature and Engineered Materials

Structural Dimension	Materials in Nature	Engineering Materials
Building Blocks	Organic compounds such as carbon, hydrogen, oxygen, nitrogen, sulfur etc.	Metals and synthetic compounds (iron, nickel, chromium, oxygen, nitrogen, copper, etc.)
Assembly Process	Biological growth and self-assembly	Controlled fabrication and design
Scales	Multi-level hierarchy across size ranges	Consistent components at a single scale
Purpose	Evolved for tailored functionalities	Engineered for targeted functions
Lifecycle	Responsive and adaptable over time	Predesigned properties and reactivity
Durability	Self-repair and healing capabilities	Retrofitting and potential rebuilding
Origin	Derived from living organisms (plants, animals, microbes)	Manufactured or synthesized by industry
Strength	Optimized strength-to-weight ratio	Engineered for specific strength and durability
Biodegradability	Often biodegradable; natural breakdown	Typically non-biodegradable; persistent
Adaptability	Can adapt to changing environments	Fixed properties
Environmental Impact	Generally eco-friendly, low pollution	May have significant environmental impact
Cost	Cost varies based on availability and extraction methods	Cost depends on production processes and raw materials
Applications	Found in biological structures, tissues, medicine, food, bioengineering	Used in construction, transportation, electronics, consumer goods, industry

copying nature's systems into artificial devices, largely due to the immense variances in structure and function.

7.2 BIOMIMICRY PRINCIPLES AND BIO-INSPIRED DESIGN

Biomimicry stands as a captivating approach that has consistently guided humanity's pursuit of ingenious solutions. Throughout history, keen observations have allowed us to harness the extraordinary adaptations and thriving strategies of natural organisms and ecosystems as they have evolved to endure. Delving into the intricate geometries, mechanisms, and behaviours of the living world has paved the way for profound insights, inspiring transformative innovations. Biomimicry transcends conventional boundaries, inspiring breakthroughs across diverse disciplines. Architects and engineers have been empowered to revolutionize building construction and structural design through bio-inspired materials and structures. These hold the potential to create greener, more energy-efficient, sustainable, and resilient built environments. While some attempts have explored the use of bio-inspired materials, large-scale applications for construction remain relatively unexplored. The applications of bio-inspiration span across various domains, encompassing concrete composite materials, bacteria-enhanced concretes and soils, building envelopes, and the design of large structures. Several natural structural composites like seashells, bone, and wood exhibit complex hierarchical designs that improve mechanical performance and have inspired various synthetic counterparts – shell-inspired materials demonstrate enhanced fracture resistance similar to nacre and abalone shells [33, 34]; bone-inspired collagen-hydroxyapatite biocomposites serve as scaffolds for tissue engineering; wood-inspired lignocellulosic composites replicate multi-scale fibre reinforcements [35, 36]. Additionally, composites with graded variations [37] in structure and properties, as seen in bone, bamboo and seashells, have led to functionally graded synthetic composites with optimized mechanical characteristics and controlled failure modes, such as lightweight high-strength concrete [38, 39] and ceramics with fracture control. Sustainable biocomposites utilizing biopolymers, natural fibres and biowaste-derived fillers further implement bio-inspired assembly, including nanocellulose-reinforced matrices [40, 41] and silk/collagen-based multifunctional films [42]. So, it is clear that nature offers diverse composite design strategies spanning reinforced, functionally graded and green composites for applications from biomedicine to construction, by studying and implementing its sophisticated hierarchical architectures. Figure 7.1 depicts a few common sources of bio-inspiration from nature and their potential applications in building materials and structures. The promise of biomimicry embarks us on an enthralling journey through nature's design brilliance, where the distinction between natural ingenuity and human innovation blurs, revealing transformative paths toward a more sustainable and harmonious future.

7.3 KEY INSPIRATION FROM NATURAL COMPOSITES

The captivating world of biomimicry beckons us to discover the extraordinary potential hidden within natural composites. This section embarks on an exploration of these remarkable materials such as collagen, bone, silk, plant biomass, each bearing distinct characteristics that have kindled inspiration for bio-inspired design and material development.

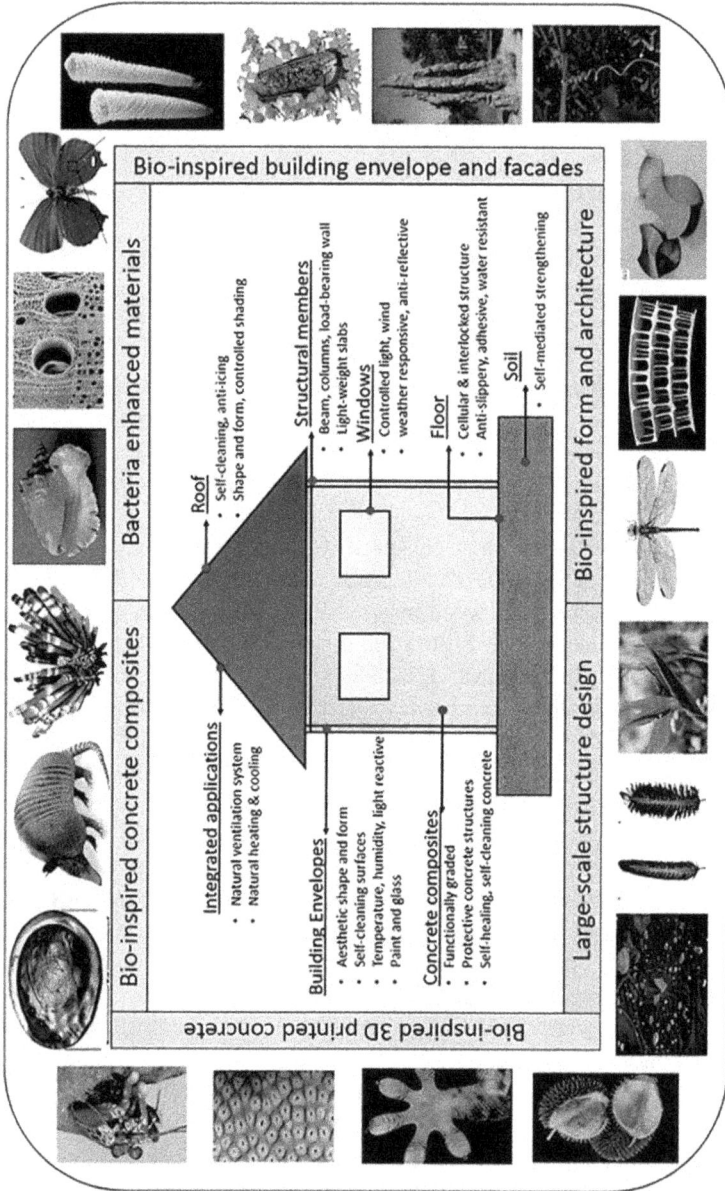

FIGURE 7.1   Exploring the possibilities of bio-inspired materials in the building and construction industry [43].

7.3.1 COLLAGEN

Collagen, a ubiquitous protein in the human body, weaves its magic as a primary component in diverse tissues like bones, teeth, tendons, and the eardrum, forming the natural extracellular matrix [44]. Of the 28 identified collagen types, Collagen Type I reigns as the most abundant, playing a pivotal role in bone structure by constituting a significant portion of the organic matrix and bone volume.

This multifaceted material boasts intriguing traits such as biodegradability, osteo-conductivity, and compatibility as a cell attachment and drug delivery carrier. Its remarkable antimicrobial properties elevate collagen's significance in crafting tissue replacements, capitalizing on advanced manufacturing techniques like direct ink writing and injection moulding.

Underpinning collagen's elegance is its distinctive hierarchical organization, depicted in Figure 7.2 [45], built upon repetitive Gly-X-Y triplets, with X and Y typically being proline and hydroxyproline. Assembling tropocollagen molecules create a fibrillary structure with an essential D-period featuring overlap and gap regions. The gap region's significance in bone tissue lies in its role as a mineral repository.

At the molecular level, a sequence of amino acid (~Å) elegantly assembles into a triplet of helix. These collagenous helices, when assembled together, form fibrils, ultimately culminating in the formation of macro tissue fibres. The beauty of collagen's architecture lies in this intricate arrangement, highlighting the seamless harmony between its molecular and macroscopic levels.

Researchers have delved into collagen's behaviour across multiple scales, unveiling its viscoelastic attributes through tests on fibres and tissues like tendons, leading to analytical models. Investigations into collagenous fibrils have revealed distinct mechanical properties depending on load application. On the tropocollagen scale, experimental and computational studies have explored collagen peptide elasticity, determining Young's modulus and viscosity [46]. Molecular dynamics models have provided insights into collagen peptide viscoelasticity, offering valuable Young's modulus, viscosity, and relaxation time data. Recent work has examined single collagen peptides under impulsive loads, exploring loading directions and hydrating effects on wave generation and energy dissipation. Interestingly, water appears to significantly impact dissipative behaviour, particularly concerning load alignment with the polymer's main axis [46].

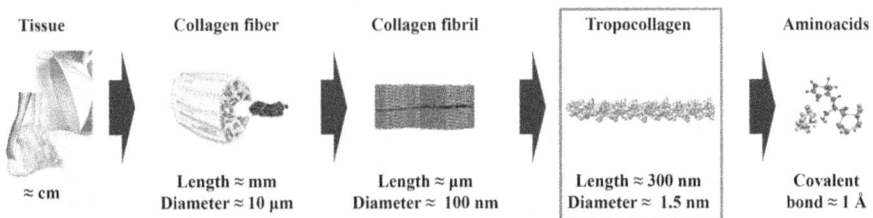

Tissue	Collagen fiber	Collagen fibril	Tropocollagen	Aminoacids
≈ cm	Length ≈ mm Diameter ≈ 10 μm	Length ≈ μm Diameter ≈ 100 nm	Length ≈ 300 nm Diameter ≈ 1.5 nm	Covalent bond ≈ 1 Å

FIGURE 7.2 The exquisite hierarchical framework of collagen [45].

7.3.2 BONE

Bone is a remarkable natural lightweight structural material that provides crucial support for diverse animal bodies across varying sizes. Fundamentally, bone can be characterized as a composite comprising a collagenous matrix reinforced by mineral platelets primarily made of calcium and phosphorus in the form of hydroxyapatite (HAP) crystals [47]. The basic building blocks of bone are HAP and collagen. HAP imparts stiffness and load-bearing strength, while collagen provides flexibility enabling bone to dissipate energy when undergoing extensive mechanical deformation.

What makes bone fascinating is the complexity of its hierarchical organization spanning seven levels of sub-structures from the atomistic to macroscale as shown in Figure 7.3 [45]. Each level has distinct arrangements tailored to serve specific functions, creating a highly optimized material [48].

At the sub-nanoscale, mineralized collagen fibrils (MCFs) constitute the basic building blocks, universally present in all bony tissues [49, 50]. At the nano-to-microscale, MCFs assemble into fibril arrays resembling composite laminae. The microscale features include lamellae, forming hollow cylindrical osteons with internal vascular canals and weak outer sheaths, characteristic of the Haversian system in cortical bone. Differentiation into trabecular and cortical tissue at larger scales allows bone to adapt its mechanical properties to diverse loads [51–54]. Apart from its remarkable structure, bone exhibits the ability to self-repair through constant remodelling – a regulated process involving resorption of old tissue and formation of new one. This remodelling process allows for structural adaptation to common loading conditions, avoiding stress concentration and facilitating post-damage repair.

Bone's multi-scale hierarchical organization leads to exceptional mechanical properties that far exceed those of its individual building blocks. Its toughness, optimal stiffness, and strength make it an attractive subject for biomimicry research. By understanding bone's hierarchical design and toughening mechanisms, scientists gain valuable insights for the design of novel synthetic materials [55]. Biomimicry, leveraging nature's diverse examples, has the potential to revolutionize material engineering, striving to match structure with function [51].

7.3.3 SILK

Silk, an extraordinary natural polymer spun by insects and arachnids (spiders), has been utilized for textiles since ancient times, with its history dating back to 27th century BC in China. However, its applications have expanded significantly due to a deeper understanding of its unique structural and biological properties. In modern times, silk

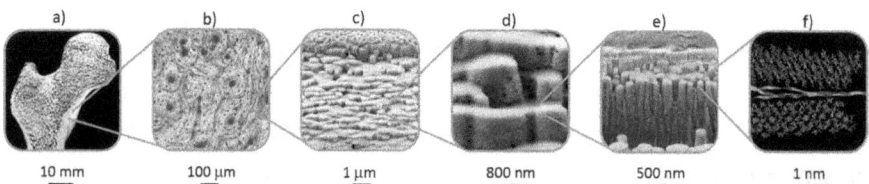

a)   b)   c)   d)   e)   f)

10 mm   100 µm   1 µm   800 nm   500 nm   1 nm

FIGURE 7.3   Bone's multi-scale hierarchy [45].

finds use not only in biotechnology, such as biocompatible hydrogels, membranes, and fibres, but also in civil engineering for sustainable filling materials [56].

The natural protein polymer silk possesses impressive mechanical properties despite its lightweight, flexible nature. With a tensile strength approaching 1.7 GPa, silk can rival conventional robust materials including steel (1.5 GPa) and Kevlar (3.6 GPa). Additionally, silk displays exceptional toughness, owing to its hierarchical protein structure [57]. This combination of strength and toughness make silk well suited for varied applications.

Interestingly, when exposed to moisture, silk undergoes a supercontraction phenomenon, shrinking in length by up to 50%. This effect causes silk to become stiffer, a behaviour leveraged by spiders to orient their webs [58].

Further adding to its intrigue, silk exhibits variability in its mechanical performance depending on factors such as spider species, silk fibre type, treatment method, age, and length. Reported statistical modulus values range from 0.6 to 7 [59]. At the nanoscale, silk consists of crystalline β-sheet regions linked together by amorphous polymer chains and stabilized via hydrogen bonds [60]. This complex hierarchical architecture allows silk to be tailored by different organisms for specialized functions. Silkworms spin protective cocoons from silk, while spider silk is engineered into prey-trapping webs [61].With its biocompatibility, exceptional mechanical performance, and the ability to undergo supercontraction, silk continues to be an exciting and promising material for future innovations in multiple industries.

7.3.4 PLANT BIOMASS

Plant biomass such as agricultural residues and grasses represent promising renewable feedstocks for bio-inspired materials. They contain biopolymers like cellulose, hemicellulose, and lignin that can be converted into fuels, chemicals, and advanced materials, adding value to the abundant biomass [62]. Wood stands out for its exceptional mechanical strength which has been leveraged in load-bearing structures since ancient times. Wood exhibits a complex multi-scale hierarchy spanning from macroscale trunks to nanoscale components. At the macrolevel, tree trunks can reach impressive heights and widths with visible growth ring patterns. Microscopically, wood has a porous structure comprising cells with multilayered walls [63]. The cell walls contain cellulose fibrils reinforcing a matrix of hemicellulose and lignin. Fibril orientations vary between the nanoscale wall layers [63]. This intricate structure provides wood's strength. Computational modelling reveals that fibril alignment and hemicellulose interactions influence biomass deformation. Further research can provide molecular insights to better utilize plant biomass as a sustainable resource for novel bio-inspired materials. Hence, the hierarchical structure of abundant, renewable plant biomass makes it a promising feedstock for developing sustainable composite materials inspired by nature's design.

7.4 BIO-INSPIRED SYNTHETIC COMPOSITES

7.4.1 BONE-LIKE COMPOSITES

Bone is a natural composite material composed of a collagen matrix reinforced with hydroxyapatite (HAP) mineral. It has a complex multi-scale hierarchical structure that provides an optimal combination of stiffness, strength, and toughness. The collagen matrix confers flexibility and fracture toughness to bone, while the HAP mineral provides stiffness and load-bearing strength [64]. Researchers have been inspired by bone to develop synthetic bone substitutes using biodegradable polymers like polycaprolactone (PCL) and polylactic acid (PLA) as the matrix, reinforced with HAP particles or fibres. These composites aim to mimic the mechanical properties of natural bone while also being biocompatible and osteoconductive to promote bone cell growth. Various additive manufacturing techniques like fused deposition modelling, inkjet 3D printing and extrusion bioprinting have been used to fabricate porous scaffolds of the PCL/HAP and PLA/HAP composites for bone tissue engineering applications. The porosity allows bone cell infiltration, vascularization and tissue in-growth. The scaffolds have exhibited mechanical properties approaching that of natural cartilage and bone, with compressive moduli in the range of 10–70 MPa [65]. In vitro studies have shown enhanced osteoblast adhesion and proliferation on the composites. The advantages of additive manufacturing allow fabricating customized implants and scaffolds with controlled architecture tailored to the defects in the bone. Future challenges include improving the mechanical properties closer to cortical bone and long-term in vivo evaluation of bone regeneration efficacy [66].

7.4.2 BIOMASS-BASED COMPOSITES

Lignocellulosic biomass from agricultural residues and grasses represents an abundant renewable resource. The biopolymers cellulose, hemicellulose and lignin derived from biomass can be utilized to fabricate fully bio-based and biodegradable composites. Cellulose fibre reinforcement has been used traditionally in composites with plastics like polypropylene. Recent interest has been in using polylactic acid (PLA) as the polymer matrix along with natural fibres to develop completely biobased composites. However, issues like high cost of PLA synthesis and fibre variability have hindered large-scale adoption. An emerging area is using lignin along with cellulose to create composites. Lignin has been considered unsuitable earlier due to its hydrophobic nature. But recent studies have shown that lignin can improve mechanical strength in cellulose-lignin films and fibres by enhancing interfacial adhesion. Such composites have also exhibited functionalities like selective gas permeability, enzyme immobilization, and photonic properties.

Moreover, further research is focused on addressing challenges in processing of cellulose-lignin composites for consistent properties and exploring their applications. Molecular modelling approaches can provide insights into the nanoscale interactions and deformation mechanisms in these materials.

7.4.3 BIOWASTE-BASED COMPOSITES

Biowaste from sources like farms, wastewater treatment plants, and industry contains valuable compounds that can be transformed into useful composite materials. Directly incorporating waste into polymers faces challenges like variability and hygiene concerns. Conversion processes are better suited to derive uniform value-added products. Hydrothermal processing uses heat and pressure with water to carbonize and decompose waste biomass. The hydrothermal biocarbons can reinforce polymers like polyvinyl alcohol to improve moisture resistance and strength of composites. The hydrochars can also be utilized directly in applications like energy storage, catalysis, and agriculture. Hydrothermal treatment is suitable for high moisture waste and avoids energy-intensive drying. Pyrolysis also carbonizes waste biomass but requires drying and substantial energy input. It produces bio-oil and biochar which can make activated carbon or polymer composites with enhanced mechanical properties and stability. Computational modelling provides insights into the molecular interactions and interfaces in biowaste composites. Experimental investigation of scalable fabrication methods and performance validation is important for translating biowaste-based composites from lab research to practical applications. Overall, conversion processes open up opportunities for developing sustainable composites by utilizing abundant and renewable biowaste feedstocks.

7.4.4 Bio-Based Electrodes

Electrodes are key components in batteries for energy storage. Bio-based carbon materials derived from sustainable sources have gained interest as alternatives to conventional electrode materials to promote sustainability. Traditionally, pyrolysis has been used to carbonize biomass waste and produce activated carbon electrodes. Hydrothermal carbonization of plant biomass also yields carbon nanofibers, graphene and conductive carbon fillers to reinforce polymer composites for electrode fabrication. Silk biopolymer composites containing conductive carbon fillers show promise as flexible electrode substrates. Biocarbons from chitin and lignin can potentially improve interfacial adhesion and conductivity compared to carbon nanotubes in the silk matrix. Computational studies reveal functional groups in biocarbons can enhance filler interaction with silk. Lignin has been directly converted into carbon fibre mats via electrospinning and carbonization for freestanding electrodes. The lignin-based electrodes exhibited higher activity compared to commercial carbon electrodes owing to their high surface area and oxygen functional groups [67].

All in all, biowaste and sustainable biopolymers offer renewable alternatives to produce carbon materials for both composite bio-electrodes and standalone electrodes. Further research is required to scale up bio-based electrodes and evaluate their real-world performance in energy storage devices.

7.5 EMERGING BIO-INSPIRED COMPOSITES

7.5.1 Quantum Dots-Polymer Nanocomposites

Quantum dots-polymer nanocomposites are an emerging class of bio-inspired composites that leverage the unique optical and electronic properties of quantum dots (QDs). QDs are semiconductor nanocrystals that exhibit quantum confinement effects, resulting in useful photonic and optoelectronic capabilities. Integration of QDs into polymer matrices creates a multifunctional nanocomposite material. For instance, encapsulation of QDs into polymer nanofibers via electrospinning produces fluorescent composite fibres with applications in biological imaging and sensing.

The QD-embedded nanofibers have shown stable fluorescence over months, making them suitable for long-term biosensing [68]. QDs have also been incorporated into graphene heterostructures to develop photocatalytic composites for solar energy harvesting. Cerium oxide QDs effectively absorb visible light and promote charge transfer in the graphene composite photocatalyst. Looking forward, the combination of QDs with biopolymers like silk presents opportunities to develop biocompatible, photonic nanocomposites [69]. The tunable optical properties and biocompatibility would be beneficial for implantable biological sensors and tissue engineering scaffolds.

7.5.2 SHAPE MEMORY BIO-INSPIRED COMPOSITES

Shape memory polymers represent an emerging category of smart materials that can transition between two distinct shape states in response to external triggers like heat, light, or moisture. Bio-inspired shape memory composites seek to imitate the actuation behaviours observed in nature, such as the coil-to-stretch motion of plant tendrils and seed pods as they react to changes in humidity. For instance, researchers synthesized a moisture-responsive graphene-polymer composite film that exhibited biomimetic coiling and uncoiling reminiscent of climbing tendrils [70]. The notable tight-coiling effect displayed by these bio-inspired humidity-sensitive films renders them promising for water harvesting from fog. More broadly, the ability to rationally engineer shape memory films with tunable deformation and recovery characteristics opens up novel possibilities for soft actuators, biomedical devices, flexible electronics, and other applications.

7.5.3 SELF-HEALING BIO-INSPIRED COMPOSITES

Self-healing materials are bio-inspired composites containing microcapsules of healing agents that allow autonomous repair of damage, thereby mimicking the self-healing capability of biological tissues. When cracks propagate through the composite matrix, rupture of the microcapsules releases the healing agent into the crack plane by capillary action. Polymerization of the healing agent bonds and closes the cracks, recovering mechanical strength. For instance, a bio-inspired self-healing polymer film demonstrated high tensile strength (34 MPa) and repeatable closure of cracks up to 200 microns through multiple damage-heal cycles [71]. The self-healing films showcase durable, flaw-tolerant behaviour desired for substrates in flexible electronics. In summary, these emerging biomimetic composites exemplify the diverse multifunctional properties and dynamic responses that can be realized through bio-inspired design strategies and synthetic material assembly.

7.6 HIERARCHICAL DESIGN VIA MACHINE LEARNING

Machine learning (ML) encompasses computational techniques that can learn and improve from data to make decisions, without the need for explicit programming. ML focuses on mapping inputs to outputs, either using hand-engineered features or directly from raw data like images. Deep learning, a type of ML, leverages neural networks to learn hierarchical representations and patterns straight from unstructured data including images, audio and text. In recent times, progress in algorithms, computing power and big data availability has enabled major advances in ML for processing complex data across domains [25, 72]. This has opened up opportunities to apply ML to long-standing materials science challenges. ML has shown promise in efficiently solving previously intractable problems like composites optimization. It can be combined with simulation tools like finite element analysis to enhance training and validate predictions. ML also holds potential for inverse materials design – predicting compositions and processing methods from target properties by learning from extensive experimental data; this can speed up discovery of new materials [72]. Overall, ML is advancing computational modelling of materials by integrating data-driven techniques across length scales - from continuum level modelling to atomistic simulations. For instance, ML-based coarse-grained models can efficiently predict water's molecular behaviour. Similarly, 'smart' finite elements leverage ML to bypass intensive physics calculations. Various optimization methods [26, 73–77] are employed to identify and tune the ideal processing conditions for optimizing the resulting composite's performance. The complex hierarchical structures found in natural materials like bone are responsible for their superior mechanical properties. Researchers across disciplines have looked to natural materials for inspiration to solve engineering problems and achieve enhanced properties in synthetic materials. This biomimicry approach implements natural design features into man-made materials. However, the sophisticated hierarchies in nature result from prolonged evolution and self-assembly processes that are challenging to replicate exactly in synthetic materials. Additive manufacturing provides more precision in recreating complex bio-inspired structures compared to conventional manufacturing. But these mimicries remain static approximations of dynamic natural materials like bone that remodel and adapt. To tailor biomimetic prototypes for specific engineering needs, optimization algorithms have been incorporated into the design process. But mimicking nature's vast design space and evolution timeframe with computers remains constrained. In recent times, ML has shown promise in materials design and discovery by learning from data. For instance, ML coupled with simulation tools has efficiently designed composites with properties comparing or exceeding bio-inspired ones. ML also offers an alternative to coarse-graining by studying systems without fully characterizing microstructures. ML models have been used to map the hierarchical structure of proteins into an analogue music space, and generate new protein sequences from musical compositions. In a recent study, a ML model known as RNN, capable of understanding the hierarchical structures present in protein sequences represented in a musical space. Remarkably, this ML model can generate entirely new protein designs by translating musical compositions back into material sequences. Their groundbreaking research offers insight into a fresh perspective on hierarchical design, highlighting how ML models can uncover the connections between diverse hierarchical systems as depicted in Figure 7.4 [45]. This demonstrates ML's potential for hierarchical design by discovering connections between complex systems.

Graph-based ML models have also been leveraged as bridges between material systems across length scales. Overall, ML brings a new data-driven approach to accelerate the design of hierarchically structured materials, complementing bio-inspiration and

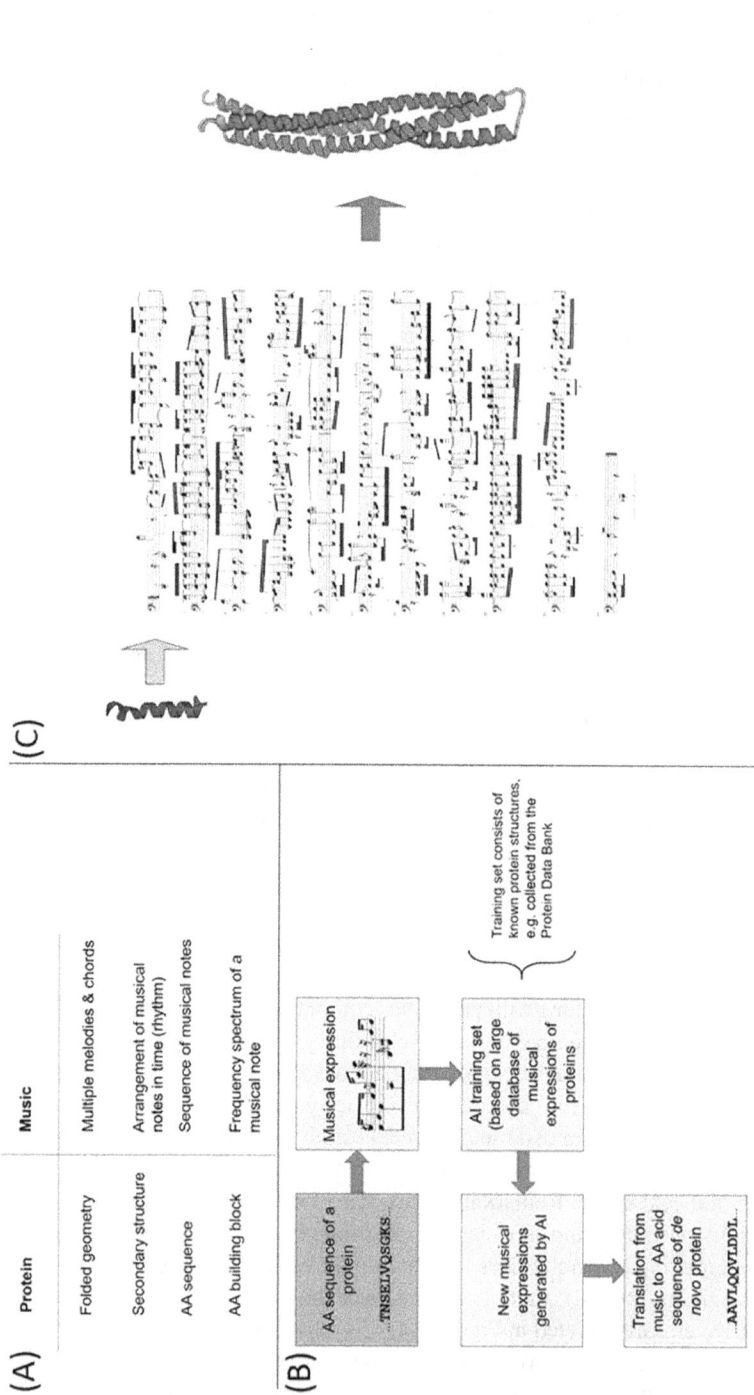

FIGURE 7.4　Illustration of the relationship between the hierarchical structure of proteins and music [45]. (a) Presents the analogy between these two hierarchical systems, (b) Demonstrates the overall flow chart of the process, (c) Shows an example of de novo protein designs achieved through the translation between protein and musical spaces using ML models.

overcoming some of its limitations. More research at the intersection of ML and biomimicry can uncover new possibilities for engineering advanced materials.

7.7 ADDITIVE MANUFACTURING AND FABRICATION

Additive manufacturing has proven to be a highly adaptable and efficient technique for fabricating intricate and complex structures, outperforming conventional methods in both precision and control. Researchers leveraged this technology to create an artificial model of a spider web, revealing insights into the strength and durability arising from the web's natural design. Additionally, additive manufacturing facilitates the synthesis of diverse natural and bio-inspired composites with fine-tuned, elaborate architectures. Its strength lies in enabling rapid fabrication and testing of composites with original material combinations and structural configurations, often supplemented with computational simulations and ML. Recent studies showcase the value of integrating computational modelling with additive manufacturing to investigate deformation and fractures in composites. Dimas and Buehler utilized simulations to digitally model the behaviour of composites with varying material distributions, while leveraging additive manufacturing to physically produce the simulated structures for experimental validation. Building on this approach, Gu et al. [78] implemented an advanced algorithm to examine how placements of strong and weak components affect overall properties. The composites were efficiently fabricated through additive manufacturing and experimentally tested to verify the computational predictions. In a follow-up study, they enhanced the methodology by incorporating ML to guide the design of composites with hierarchical architecture. Pushing the boundaries further, Milazzo, Buehler and team recently employed additive manufacturing to develop hydroxyapatite-silk composites tailored as tissue engineering scaffolds. The concerted integration of modelling, simulation, ML and advanced digital fabrication portends new possibilities in expediting materials development across scientific realms.

Overall, these findings highlight the synergistic potential of combining computational simulations, ML algorithms and additive manufacturing to accelerate bio-inspired materials discovery and optimization in a rapid yet controlled manner.

7.8 CONCLUSION

Nature's genius has served as an unlimited source of inspiration for human innovation across diverse fields. This chapter provided an overview of core concepts and principles of biomimicry, highlighting natural composites like collagen, bone, silk, and plant biomass as key sources of bio-inspiration. Their unique multi-scale hierarchical structures and mechanical properties provide blueprints for engineering advanced synthetic composites. Additive manufacturing presents exciting possibilities for recreating the complexity of natural materials. ML also offers new avenues for optimized hierarchical design by discovering connections across systems. However, limitations persist in replicating the dynamic features of biological materials like self-healing capabilities. Advanced characterization techniques from molecular to macroscale continue to uncover structure–property relationships in natural composites, providing insights to guide materials development. Significant progress has been made in fabrication of bone-like, biomass-based, and biowaste-derived composites. Emerging areas like quantum dot-polymer nanocomposites, shape memory, and self-healing bio-inspired composites demonstrate the potential for multifunctionality. Looking ahead, integration of biomimicry principles with computational modelling, manufacturing, and testing will enable further discoveries. Opportunities exist in designing smart composites with sensing abilities, controlled degradation and greater sustainability. Hierarchical composite architectures can also be tailored to meet specific engineering needs.

Overall, bio-inspiration remains a compelling guiding philosophy for future materials research and development. By blending nature's time-tested genius with human innovation, scientists can continue pushing boundaries in designing high-performance composites that balance optimal function with environmental harmony. The synergies between biology, engineering and emerging technologies illuminated in this chapter chart promising directions in that endeavour.

REFERENCES

[1] N. Ranjan, R. Tyagi, R. Kumar and A. Babbar, 3D printing applications of thermo-responsive functional materials: A review, *Advances in Materials and Processing Technologies* 11 (2023), pp. 1–17.

[2] A.S. Uppal, A. Sharma, A. Babbar, K. Singh and A.K. Singh, Minimum quality lubricant (MQL) for ultraprecision machining of titanium nitride-coated carbide inserts: Sustainable Manufacturing process, *International Journal on Interactive Design and Manufacturing (IJIDeM)* (2023), pp. 1–12. https://doi.org/10.1007/s12008-023-01299-4

[3] A. Babbar, V. Jain, D. Gupta, K. Goyal, C. Prakash, K. Saxena et al., Investigation of infrared thermography of cortical bone grinding in neurosurgery, *Advances in Science and Technology Research Journal* 17 (2023), pp. 116–123.

[4] S. Bansal, S. Kaushal, J. Mago, D. Gupta, V. Jain, A. Babbar et al., Effect of variation of WC reinforcement on metallurgical and cavitation erosion behavior of microwave processed NiCrSiC-WC composites clads, *Proceedings of the Institution of Mechanical Engineers, Part C: Journal of Mechanical Engineering Science* 237 (2023), pp. 5460–5475.

[5] A. Kumar, A. Babbar, V. Jain, D. Gupta, B.P. Pathri, C. Prakash et al., Investigation and enhancement of mechanical properties of SS-316 weldment using TiO_2-SiO_2-Al_2O_3 hybrid flux, *International Journal on Interactive Design and Manufacturing (IJIDeM)* (2023). https://doi.org/0.1007/s12008-023-01198-8

[6] Y. Tian, C. Tian, J. Han, A. Babbar and B. Liu, Characteristics of grinding force and Kevlar deformation of novel body-armor-like abrasive tool, *The International Journal of Advanced Manufacturing Technology* 122 (2022), pp. 2019–2030.

[7] Z. Gu, Y. Tian, J. Han, C. Wei, A. Babbar and B. Liu, Characteristics of high-shear and low-pressure grinding for Inconel718 alloy with a novel super elastic composite abrasive tool, *The International Journal of Advanced Manufacturing Technology* 123 (2022), pp. 345–355.

[8] A. Babbar, A. Sharma, V. Jain and D. Gupta, *Additive Manufacturing Processes in Biomedical Engineering*, CRC Press, Boca Raton, 2022.

[9] V. Kumar, C. Prakash, A. Babbar, S. Choudhary, A. Sharma and A.S. Uppal, Additive Manufacturing in Biomedical Engineering, in Atul Babbar, Ankit Sharma, Vivek Jain, Dheeraj Gupta (Eds.), *Additive Manufacturing Processes in Biomedical Engineering*, CRC Press, Boca Raton, 2022, pp. 143–164.

[10] B.P. Pathri, Mohd S. Khan and A. Babbar, Relevance of Bio-Inks for 3D Bioprinting, in Atul Babbar, Ankit Sharma, Vivek Jain, Dheeraj Gupta (Eds.), *Additive Manufacturing Processes in Biomedical Engineering*, CRC Press, Boca Raton, 2022, pp. 81–98.

[11] A. Babbar, V. Jain, D. Gupta, A. Sharma, C. Prakash, V. Kumar et al., Additive Manufacturing for the Development of Biological Implants, Scaffolds, and Prosthetics, in Atul Babbar, Ankit Sharma, Vivek Jain, Dheeraj Gupta (Eds.), *Additive Manufacturing Processes in Biomedical Engineering*, CRC Press, Boca Raton, 2022, pp. 27–46.

[12] A. Babbar, Y. Tian, V. Kumar and A. Sharma, *3D Bioprinting in Biomedical Applications*, in *Additive Manufacturing of Polymers for Tissue Engineering*, CRC Press, Boca Raton, 2022, pp. 1–16.

[13] A. Sharma, A. Babbar, Y. Tian, B.P. Pathri, M. Gupta and R. Singh, Machining of ceramic materials: A state-of-the-art review, *International Journal on Interactive Design and Manufacturing (IJIDeM)* 17 (2022), pp. 2891–2911.

[14] A. Babbar, V. Jain, D. Gupta and D. Agrawal, Histological evaluation of thermal damage to Osteocytes: A comparative study of conventional and ultrasonic-assisted bone grinding, *Medical Engineering & Physics* 90 (2021), pp. 1–8.

[15] A. Babbar, V. Jain, D. Gupta, C. Prakash and D. Agrawal, *Potential Application of CEM43°C and Arrhenius Model in Neurosurgical Bone Grinding*, Springer, Cham, 2022, pp. 145–158.

[16] A. Babbar, V. Jain, D. Gupta, D. Agrawal, C. Prakash, S. Singh et al., Experimental analysis of wear and multi-shape burr loading during neurosurgical bone grinding, *Journal of Materials Research and Technology* 12 (2021), pp. 15–28.

[17] A. Babbar, V. Jain, D. Gupta and D. Agrawal, Finite element simulation and integration of CEM43°C and Arrhenius Models for ultrasonic-assisted skull bone grinding: A thermal dose model, *Medical Engineering & Physics* 90 (2021), pp. 9–22.

[18] A. Babbar, V. Jain and D. Gupta, In vivo evaluation of machining forces, torque, and bone quality during skull bone grinding, *Proceedings of the Institution of Mechanical Engineers, Part H: Journal of Engineering in Medicine* 234 (2020), pp. 626–638.

[19] A. Babbar, V. Jain and D. Gupta, Thermogenesis mitigation using ultrasonic actuation during bone grinding: A hybrid approach using CEM43°C and Arrhenius model, *Journal of the Brazilian Society of Mechanical Sciences and Engineering* 41 (2019), p. 401.

[20] A. Babbar, A. Sharma and P. Singh, Multi-objective optimization of magnetic abrasive finishing using grey relational analysis, *Materials Today: Proceedings* 50 (2022), pp. 570–575.

[21] A. Sharma, M. Kalsia, A.S. Uppal, A. Babbar and V. Dhawan, Machining of hard and brittle materials: A comprehensive review, *Materials Today: Proceedings* 50 (2022), pp. 1048–1052.

[22] A. Babbar, A. Rai and A. Sharma, Latest trend in building construction: Three-dimensional printing, *Journal of Physics: Conference Series* 1950 (2021), p. 012007.

[23] P. Khanduja, H. Bhargave, A. Babbar, P. Pundir and A. Sharma, Development of two-dimensional plotter using programmable logic controller and human machine interface, *Journal of Physics: Conference Series* 1950 (2021), p. 012012.

[24] G. Kalia, A. Sharma and A. Babbar, Use of three-dimensional printing techniques for developing biodegradable applications: A review investigation, *Materials Today: Proceedings* 62 (2022), pp. 346–352.

[25] C. Prakash, V. Kumar, A. Mistri, A.S. Uppal, A. Babbar, B.P. Pathri et al., Investigation of functionally graded adherents on failure of socket joint of FRP composite tubes, *Materials* 14 (2021), p. 6365.

[26] A. Sharma, V. Kumar, A. Babbar, V. Dhawan, K. Kotecha and C. Prakash, Experimental investigation and optimization of electric discharge machining process parameters using Grey-fuzzy-based hybrid techniques, *Materials* 14 (2021), p. 5820.

[27] A. Babbar, V. Jain, D. Gupta and C. Prakash, Experimental investigation and parametric optimization of neurosurgical bone grinding under bio-mimic environment, *Surface Review and Letters* 30 (2023), p. 2141005.

[28] A. Babbar, A. Sharma, R. Kumar, P. Pundir and V. Dhiman, Functionalized biomaterials for 3D printing: An overview of the literature, in Sunpreet Singh, Chaudhery Mustansar Hussain (Eds.), *Additive Manufacturing with Functionalized Nanomaterials*, Elsevier, 2021, pp. 87–107.

[29] G. Singh, A. Babbar, V. Jain and D. Gupta, Comparative statement for diametric delamination in drilling of cortical bone with conventional and ultrasonic assisted drilling techniques, *Journal of Orthopaedics* 25 (2021), pp. 53–58.

[30] M.S. Aziz and A.Y. El Sherif, Biomimicry as an approach for bio-inspired structure with the aid of computation, *Alexandria Engineering Journal* 55 (2016), pp. 707–714.

[31] J.F. V. Vincent, Stealing Ideas from Nature, in S. Pellegrino (Ed.), *Deployable Structures*, 2001, pp. 51–58. Vienna: Springer. https://doi.org/10.1007/978-3-7091-2584-7

[32] H. Schmitz, H. Soltner and H. Bousack, Biomimetic infrared sensors based on photomechanic infrared receptors in pyrophilous (fire-loving) insects, *IEEE Sensors Journal* 12 (2012), pp. 281–288.

[33] D.G. Soltan, R. Ranade and V.C. Li, A bio-inspired cementitious composite for high energy absorption in infrastructure applications, In 13th International Symposium on Multiscale, Multifunctional and Functionally Graded Materials 2014, pp. 1–4.

[34] D.G. Soltan and V.C. Li, Nacre-inspired composite design approaches for large-scale cementitious members and structures, *Cement and Concrete Composites* 88 (2018), pp. 172–186.

[35] V. Nguyen-Van, P. Tran, C. Peng, L. Pham, G. Zhang and H. Nguyen-Xuan, Bioinspired cellular cementitious structures for prefabricated construction: Hybrid design & performance evaluations, *Automation in Construction* 119 (2020), p. 103324.

[36] I.H. Chen, J.H. Kiang, V. Correa, M.I. Lopez, P.Y. Chen, J. McKittrick et al., Armadillo armor: Mechanical testing and micro-structural evaluation, *Journal of the Mechanical Behavior of Biomedical Materials* 4 (2011), pp. 713–722.

[37] V. Kumar, P.P. Das and S. Chakraborty, Grey-fuzzy method-based parametric analysis of abrasive water jet machining on GFRP composites, *Sādhanā* 45 (2020), p. 106.

[38] Y.L. Tee, M. Leary and P. Tran, Porcupine quill: Buckling resistance analysis and design for 3D printing, *Lecture Notes in Civil Engineering* 101 (2021), pp. 1311–1319.

[39] N. Toader, W. Sobek and K.G. Nickel, Energy absorption in functionally graded concrete bioinspired by sea urchin spines, *Journal of Bionic Engineering* 14 (2017), pp. 369–378.

[40] P. Tran and C. Peng, Triply periodic minimal surfaces sandwich structures subjected to shock impact, *Journal of Sandwich Structures and Materials* 23 (2021), pp. 2146–2175.

[41] C. Peng and P. Tran, Bioinspired functionally graded gyroid sandwich panel subjected to impulsive loadings, *Composites Part B: Engineering* 188 (2020), p. 107773.

[42] Y. Zhang, H. Yao, C. Ortiz, J. Xu and M. Dao, Bio-inspired interfacial strengthening strategy through geometrically interlocking designs, *Journal of the Mechanical Behavior of Biomedical Materials* 15 (2012), pp. 70–77.

[43] Ahamed, Mohammad Kaiser, Hongxu, Wang and Hazell, Paul J., From biology to biomimicry: Using nature to build better structures – A review. *Construction and Building Materials* 30 (2022), p. 126195.

[44] X.B. Yang, R.S. Bhatnagar, S. Li and R.O.C. Oreffo, Biomimetic collagen scaffolds for human bone cell growth and differentiation, *Tissue Engineering* 10 (2004), pp. 1148–1159.

[45] M. Milazzo, F. Libonati, S. Zhou, K. Guo and M.J. Buehler, Biomimicry for Natural and Synthetic Composites and Use of Machine Learning in Hierarchical Design, in *Biomimicry for Materials, Design and Habitats: Innovations and Applications*, 2022, pp. 141–182. https://doi.org/10.1016/C2019-0-01384-2

[46] A. Gautieri, S. Vesentini, A. Redaelli and M.J. Buehler, Viscoelastic properties of model segments of collagen molecules, *Matrix Biology* 31 (2012), pp. 141–149.

[47] K. Piekarski, Analysis of bone as a composite material, *International Journal of Engineering Science* 11 (1973), pp. 557–558.

[48] S. Weiner and H.D. Wagner, The material bone: Structure-mechanical function relations, *Annual Review of Materials Science* 28 (1998), pp. 271–298.

[49] T. Ackbarow and M.J. Buehler, Hierarchical coexistence of universality and diversity controls robustness and multi-functionality in protein materials, *Journal of Computational and Theoretical Nanoscience* 5 (2008), pp. 1193–1204.

[50] I. Jäger and P. Fratzl, Mineralized collagen fibrils: A mechanical model with a staggered arrangement of mineral particles, *Biophysical Journal* 79 (2000), pp. 1737–1746.

[51] M.J. Mirzaali, V. Mussi, P. Vena, F. Libonati, L. Vergani and M. Strano, Mimicking the loading adaptation of bone microstructure with aluminum foams, *Materials & Design* 126 (2017), pp. 207–218.

[52] C.H. Turner, *Three rules for bone adaptation to mechanical stimuli, Bone* 23 (1998), pp. 399–407.

[53] R. Weinkamer and P. Fratzl, Mechanical adaptation of biological materials – The examples of bone and wood, in *Materials Science and Engineering C* 31 (2011), pp. 1164–1173.

[54] F. Libonati and L. Vergani, Understanding the structure-property relationship in cortical bone to design a biomimetic composite, *Composite Structures* 139 (2016), pp. 188–198.

[55] M.E. Launey, M.J. Buehler and R.O. Ritchie, On the mechanistic origins of toughness in bone, *Annual Review of Materials Research* 40 (2010), pp. 25–53.

[56] R.M. Reddy, Innovative and multidirectional application of natural fibre, silk – A review, *Academic Journal of Entomology* 2 (2009), pp. 71–75.

[57] I. Agnarsson, M. Kuntner and T.A. Blackledge, Bioprospecting finds the toughest biological material: Extraordinary silk from a giant riverine orb spider, *PLoS One* 5 (2010), pp. 1–8.

[58] T. Giesa, R. Schuetz, P. Fratzl, M.J. Buehler and A. Masic, Unraveling the molecular requirements for macroscopic silk supercontraction, *ACS Nano* 11 (2017), pp. 9750–9758.

[59] P. Colomban, H.M. Dinh, A. Bunsell and B. Mauchamp, Origin of the variability of the mechanical properties of silk fibres: 1 - The relationship between disorder, hydration and stress/strain behaviour, *Journal of Raman Spectroscopy* 43 (2012), pp. 425–432.

[60] S. Keten and M.J. Buehler, Atomistic model of the spider silk nanostructure, *Applied Physics Letters* 96 (2010), p. 153701.

[61] Z. Qin, B.G. Compton, J.A. Lewis and M.J. Buehler, Structural optimization of 3D-printed synthetic spider webs for high strength, *Nature Communications* 6 (2015), p. 7038.

[62] B.M.H.J. Langholtz, L.M. Stokes and Eaton, 2016 Billion-ton report advancing domestic resources for a thriving bioeconomy, Oak Ridge National Laboratory 1160 (2016), pp. 448–2172.

[63] K. Jin, Z. Qin and M.J. Buehler, Molecular deformation mechanisms of the wood cell wall material, *Journal of the Mechanical Behavior of Biomedical Materials* 42 (2015), pp. 198–206.

[64] M. Milazzo, S. Danti, F. Inglese, G. Jansen van Vuuren, V. Gramigna, G. Bonsignori et al., Ossicular replacement prostheses from banked bone with ergonomic and functional geometry, *Journal of Biomedical Materials Research* 105 (2017), pp. 2495–2506.

[65] T. Huang, C. Fan, M. Zhu, Y. Zhu, W. Zhang and L. Li, 3D-printed scaffolds of biomineralized hydroxyapatite nanocomposite on silk fibroin for improving bone regeneration, *Applied Surface Science* 467–468 (2019), pp. 345–353.

[66] L. Sun, S.T. Parker, D. Syoji, X. Wang, J.A. Lewis and D.L. Kaplan, Direct-write assembly of 3D silk/hydroxyapatite scaffolds for bone co-cultures, *Advanced Healthcare Materials* 1 (2012), pp. 729–735.

[67] Y.H. Ro, J.F. Hunt and R.E. Rowlands, Technical note: Stress analysis of cellulosic-manure composites, *Wood and Fiber Science* 49 (2017), pp. 231–233.

[68] B. Dhandayuthapani, A.C. Poulose, Y. Nagaoka, T. Hasumura, Y. Yoshida, T. Maekawa et al., Biomimetic smart nanocomposite: In vitro biological evaluation of zein electrospun fluorescent nanofiber encapsulated CdS quantum dots, *Biofabrication* 4 (2012), p. 025008.

[69] J. Qian, Y. Cao, Z. Chen, C. Liu and X. Lu, Biomimetic synthesis of cerium oxide nanosquares on RGO and their enhanced photocatalytic activities, *Dalton Transactions* 46 (2017), pp. 547–553.

[70] S.M. Felton, M.T. Tolley, B. Shin, C.D. Onal, E.D. Demaine, D. Rus et al., Self-folding with shape memory composites, *Soft Matter* 9 (2013), pp. 7688–7694.

[71] Nan Zhong and Wouter Post, Self-repair of structural and functional composites with intrinsically self-healing polymer matrices: A review, *Composites Part A: Applied Science and Manufacturing* 69 (2015), pp. 226–239.

[72] D. Mayerich, R. Sun and J. Guo, Deep Learning, in *Microscope Image Processing*, Second Edition, 2022, pp. 431–456.

[73] S. Chakraborty, P.P. Das and V. Kumar, Application of grey-fuzzy logic technique for parametric optimization of non-traditional machining processes, *Grey Systems: Theory and Application* 8 (2018), pp. 46–68.

[74] V. Kumar, S. Diyaley and S. Chakraborty, Teaching-learning-based parametric optimization of an electrical discharge machining process, *Facta Universitatis, Series: Mechanical Engineering* 18 (2020), pp. 281–300.

[75] S. Chakraborty and V. Kumar, Development of an intelligent decision model for non-traditional machining processes, *Decision Making: Applications in Management and Engineering* 4 (2021), pp. 194–214.

[76] V. Kumar and S. Chakraborty, Analysis of the Surface Roughness Characteristics of EDMed Components Using GRA Method, in Proceedings of the International Conference on Industrial and Manufacturing Systems (CIMS-2020) 2022, pp. 461–478.

[77] V. Kumar, A. Mohata, A. Mistri and M. Bartoszuk, Application of SWARA-CoCoSo-based approach for tool selection of an electrical discharge machining process, *Sustainable Production, Instrumentation and Engineering Sciences* 1 (2023), pp. 19–26.

[78] G.X. Gu, C.T. Chen, D.J. Richmond and M.J. Buehler, Bioinspired hierarchical composite design using machine learning: Simulation, additive manufacturing, and experiment, *Materials Horizons* 5 (2018), pp. 939–945.

8 Machining of Carbon Nanotubes Reinforced Epoxy Composites

Harpreet Kaur Channi
Chandigarh University, Mohali, India

Raman Kumar
Guru Nanak Dev Engineering College, Ludhiana, India

Ramandeep Singh Sidhu
Guru Nanak Dev Engineering College, Ludhiana, Punjab, India

Rajender Kumar
Manav Rachna International Institute of Research & Studies, Faridabad, India

8.1 INTRODUCTION

Carbon nanotubes (CNTs) have gained significant attention recently due to their exceptional mechanical, electrical, and thermal properties. These unique characteristics make them promising reinforcements for polymer composites, especially in epoxy matrices, resulting in the development of CNT-reinforced epoxy composites with enhanced performance and multi-functionality. However, fully utilizing these advanced materials in engineering applications relies on their outstanding properties and the ability to fabricate complex shapes and precise components. This critical aspect emphasizes the significance of studying the machining of CNT-reinforced epoxy composites [1].

Machining is a fundamental manufacturing process that involves material removal from a workpiece using various cutting tools and techniques [2–28]. However, machining CNT-reinforced epoxy composites presents numerous challenges due to the heterogeneity and anisotropic nature of the composite structure, as well as the mechanical properties of CNTs themselves. The machining process must be optimized to preserve the unique properties of both the CNTs and the epoxy matrix while achieving desired dimensional accuracy and surface quality [29]. In this context, this research aims to delve into CNT-reinforced epoxy composites' machining aspects, explore traditional and advanced machining techniques, understand the effects of

DOI: 10.1201/9781003427735-8

cutting parameters, and address the associated difficulties. The insights gained from this study will contribute to a comprehensive understanding of the machinability of these composites, opening doors to their broader practical applications in industries such as aerospace, automotive, electronics, and more. By overcoming the machining challenges, engineers and researchers can fully harness the potential of CNT-reinforced epoxy composites, creating innovative and high-performance components for the next generation of advanced engineering applications [30].

CNTs are cylindrical nanostructures composed of carbon atoms arranged in a hexagonal lattice. They exhibit remarkable mechanical, electrical, and thermal properties, including high tensile strength, excellent electrical conductivity, and exceptional thermal conductivity. These unique properties have led to extensive research and interest in incorporating CNTs as reinforcements in various materials, including polymers [31].

Epoxy composites are a class of materials that consist of epoxy resin matrices reinforced with different filler materials, such as fibres, nanoparticles, and CNTs. Adding CNTs to epoxy matrices can significantly enhance the mechanical and functional properties of the resulting composites, making them highly attractive for advanced engineering applications. One of the key challenges in the practical application of CNT-reinforced epoxy composites is the fabrication of complex shapes and components [32]. As a conventional manufacturing process, machining plays a crucial role in shaping, cutting, and finishing materials to precise dimensions and surface qualities. However, machining CNT-reinforced epoxy composites is a complex task for several reasons, as shown in Figure 8.1.

- **Anisotropic Structure**: CNT-reinforced epoxy composites often exhibit anisotropic behaviour, where the mechanical properties vary with the orientation of the CNTs. This anisotropy can lead to non-uniform material removal during machining.
- **Nanotube Agglomeration**: CNTs tend to form agglomerates within the epoxy matrix, creating local variations in material properties and complicating machining operations.
- **Tool Wear**: CNTs in the composite can cause higher tool wear due to the nanotubes' high hardness and abrasive nature.
- **Thermal Effects**: Machining CNT-reinforced epoxy composites can generate significant heat, leading to thermal degradation of the epoxy matrix and affecting the material properties [33].

Understanding the machinability of CNT-reinforced epoxy composites is crucial for their successful integration into real-world applications. Research in this area aims to explore suitable machining techniques, optimize cutting parameters, and develop effective cooling and lubrication strategies to overcome the challenges posed by these advanced materials' unique structures and properties. By addressing these issues, engineers can harness the full potential of CNT-reinforced epoxy composites and realize their benefits in various industries, ranging from aerospace and automotive to electronics and biomedical applications. The objectives are to investigate the

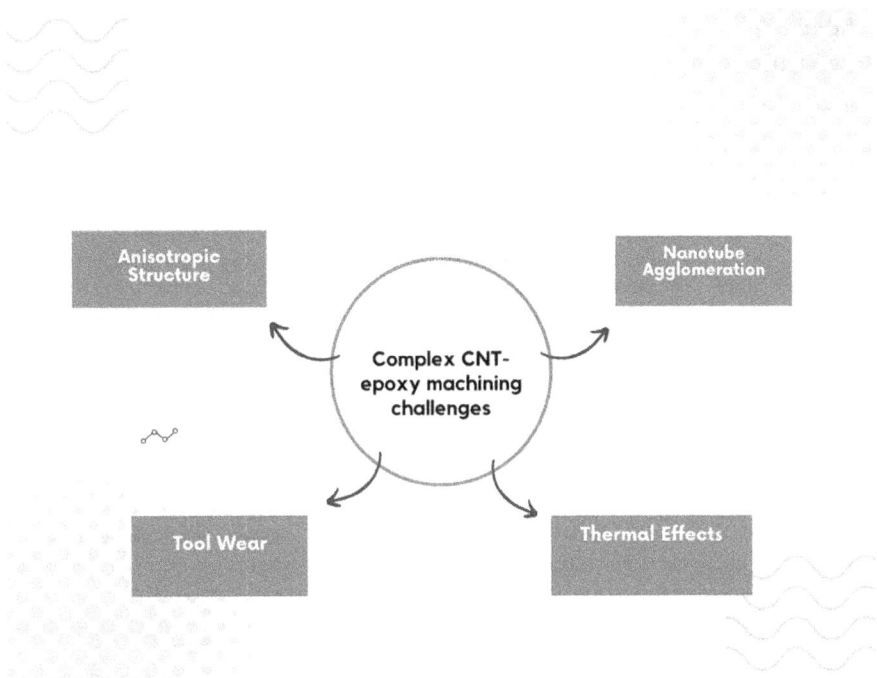

FIGURE 8.1   Complex CNT- epoxy machining challenges.

machinability of CNT-reinforced epoxy composites, optimize cutting parameters, understand tool-workpiece interaction, characterize material properties, address surface integrity, compare machining techniques, and enable practical applications in engineering fields [34–36].

8.2   LITERATURE REVIEW

The literature review on "Machining of carbon nanotube-reinforced epoxy composites" provides an overview of the existing research and studies related to the machining process of these advanced materials. It covers various aspects, including machining techniques, cutting parameters, tool materials, challenges, and the effects of machining on the microstructure and mechanical properties of CNT-reinforced epoxy composites [37]. The review begins by introducing the significance of CNT-reinforced epoxy composites in engineering applications due to their unique properties. It then explores the different types of CNTs and their methods of synthesis and dispersion within epoxy matrices [38].

Various traditional and advanced machining techniques used for machining CNT-reinforced epoxy composites are discussed, including turning, milling, drilling, abrasive water jet, laser, and electrical discharge machining. Each method's advantages, limitations, and suitability for specific applications are analysed. Studies focusing on

TABLE 8.1

Interventions and Outcomes

Ref No	Interventions	Outcomes
[40]	Engineered multiscale carbon nanotube carbon fibre reinforcement	• Interlaminar Shear Strength • Plane Electrical Conductivity
[41]	Modification of the epoxy matrix with multi-wall carbon nanotubes and a commercial block copolymer to enhance the dispersion of the nanotubes within the matrix	• Glass Transition Temperature • Compressive Strength • Interlaminar Shear Strength
[42]	Reinforcement roles of carbon nanotubes in epoxy composites with different matrix stiffness	• Reinforcement Role of Carbon Nanotubes • Mechanical Properties of Composite • Fracture Strain • Interface Interaction Between Carbon Nanotubes and Matrix
[43]	Ga/CNT epoxy nanocomposites	• Thermo Mechanical Properties • Electrical Properties • Flame Retardant Properties
[44]	Incorporating carbon nanotubes and short carbon fibres into an epoxy matrix to fabricate a high-performance multiscale composite	• Elastic Modulus • Storage Modulus • Strength • Impact Resistance
[45]	Carbon fibre reinforced epoxy composites were modified with 2 wt.%	• 3 Point Bending Test • Dynamic Mechanical Analysis (Dma) • Low Velocity Impact Test (Lvi)

optimizing cutting parameters such as cutting speed, feed rate, and depth of cut are reviewed, highlighting their influence on material removal rates, surface quality, and tool wear during machining [39]. The literature review also addresses challenges encountered during the machining process, such as tool wear, tool life, thermal effects, and potential damage to the CNT network within the composite structure.

Furthermore, the impact of machining on CNT-reinforced epoxy composites' microstructure and mechanical properties is discussed. This includes changes in dispersion, accumulation, and potential degradation of CNTs and the epoxy matrix due to the machining-induced heat. Finally, the review concludes by identifying research gaps and areas for future investigation, aiming to provide a comprehensive understanding of the state of knowledge in machining CNT-reinforced epoxy composites and guiding further advancements in this field Table 8.1 shows main interventions and outcomes.

8.3 CARBON NANOTUBES

8.3.1 STRUCTURE AND PROPERTIES OF CARBON NANOTUBES

CNTs are cylindrical nanostructures composed of carbon atoms arranged in a hexagonal lattice, forming seamless tubes with exceptional properties. The structure and properties of CNTs are key factors contributing to their widespread interest and

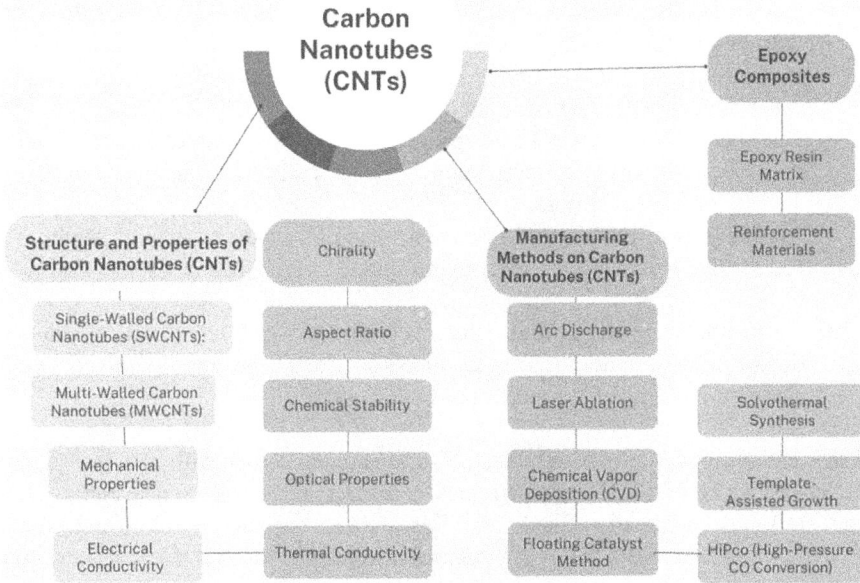

FIGURE 8.2   Overview of carbon nanotubes (CNTs).

potential applications in various fields [46] An overview of the structure and properties of CNTs is shown in Figure 8.2 and discussed below:

8.3.1.1 Single-Walled Carbon Nanotubes

Single-Walled Carbon Nanotubes (SWCNTs) consist of a single layer of carbon atoms rolled into a seamless cylindrical tube. They can be thought of as graphene sheets rolled up into a tube. The diameter of SWCNTs typically ranges from a few angstroms to a few nanometres.

8.3.1.2 Multi-Walled Carbon Nanotubes

Multi-Walled Carbon Nanotubes (MWCNTs) consist of multiple layers of carbon atoms, forming concentric tubes nested within one another. These tubes can vary in diameter, and the number of layers can range from a few to tens or even hundreds.

8.3.1.3 Mechanical Properties

CNTs possess exceptional mechanical strength and stiffness. They have one of the highest tensile strengths of any known material, making them extremely resilient to mechanical forces. Their Young's modulus is also very high, reflecting their stiffness. Additional properties include:

- **Electrical Conductivity**: CNTs exhibit excellent electrical conductivity, comparable to or even surpassing that of copper or other metals. Their unique 1D structure allows for efficient electron transport along the tube axis [47].

- **Thermal Conductivity**: CNTs have outstanding thermal conductivity, enabling efficient heat dissipation. They can conduct heat more effectively than most materials, making them valuable for thermal management applications.
- **Optical Properties**: CNTs exhibit interesting optical properties, depending on their structure and diameter. They can absorb and emit light across a wide range of wavelengths, from ultraviolet to infrared.
- **Chemical Stability**: CNTs are generally chemically stable, especially if they have high structural perfection. They are resistant to many chemicals and environmental factors.
- **Aspect Ratio**: CNTs typically have a very high aspect ratio, with lengths that can be several micrometres to millimetres, while their diameters are on the nanoscale.
- **Chirality**: CNTs can exhibit different chiralities, which refer to the orientation of the carbon hexagons in the tube's structure. The chirality affects the electronic and optical properties of CNTs [48].

These extraordinary properties make CNTs versatile materials with immense potential in various fields, including electronics, aerospace, materials science, energy storage, biomedical applications, and more. Researchers continue to explore and exploit the unique characteristics of CNTs to unlock novel applications and advance existing technologies [49].

8.4   MANUFACTURING METHODS OF CARBON NANOTUBES

CNTs can be manufactured using various methods, as shown in Figure 8.2, each with its advantages and limitations. The choice of manufacturing method depends on the desired properties of the CNTs and the intended application [50]. Some common manufacturing methods for CNTs are discussed below:

8.4.1   ARC DISCHARGE

In this method, two graphite electrodes are subjected to high electrical current in an inert gas atmosphere. The high temperature generated causes the carbon atoms to vaporize and form CNTs. This method can produce both SWCNTs and MWCNTs.

8.4.2   LASER ABLATION

A pulsed laser is used to vaporize a graphite target in the presence of a metal catalyst. The carbon atoms condense to form CNTs on the surface of the catalyst. This method is suitable for producing high-quality SWCNTs.

8.4.3   CHEMICAL VAPOR DEPOSITION

In CVD, carbon-containing gases (e.g., hydrocarbons) are decomposed in the presence of a metal catalyst at high temperatures. The carbon atoms are deposited on the catalyst surface, forming CNTs. CVD can produce both SWCNTs and MWCNTs and allows for control over the growth process.

8.4.4   FLOATING CATALYST METHOD

This method involves the injection of carbon-containing gas into a reactor along with a metal catalyst precursor. The catalyst nanoparticles and carbon feedstock react, leading to CNT growth. It is suitable for producing large quantities of MWCNTs [51].

8.4.5   HiPco (HIGH-PRESSURE CO CONVERSION)

In this method, carbon monoxide gas is converted to CNTs at high pressure in the presence of metal catalyst particles. HiPco is known for producing high-quality SWCNTs. Template-Assisted Growth: CNTs can be grown inside nanoporous templates, such as anodized aluminium oxide or zeolites. The template provides a confined space for CNT growth and allows for controlled diameter and alignment [52].

8.4.6   SOLVOTHERMAL SYNTHESIS

This method involves the reaction of carbon precursors in a high-temperature solvent. CNTs are formed through a self-assembly process, using metal catalysts or templates. Each of these manufacturing methods has specific advantages in terms of CNT quality, scalability, and control over properties. Researchers continue to explore and improve these techniques to tailor CNT properties for various applications, including electronics, nanocomposites, energy storage, sensors, and biomedical applications [53].

8.5   EPOXY COMPOSITES

Epoxy composites are a class of advanced materials consisting of epoxy resin as the matrix and one or more reinforcement materials. These composites are engineered to combine the excellent properties of epoxy resins with the enhanced mechanical, thermal, and functional characteristics of the reinforcements [54]. The resulting epoxy composites offer superior performance compared to traditional materials, making them widely used in various industries. Key components of epoxy composites:

8.5.1   EPOXY RESIN MATRIX

Epoxy resin is a type of thermosetting polymer known for its high strength, excellent adhesion, chemical resistance, and low shrinkage during curing. The epoxy resin acts as the matrix material that holds and binds the reinforcement together, providing mechanical strength and protection against environmental factors [55].

8.5.2   Reinforcement Materials

The reinforcement materials are added to the epoxy matrix to enhance specific properties of the composite. Common types of reinforcements include:

a. **Fibers**: Fiberglass, carbon fibres, aramid fibres (Kevlar), and natural fibres like hemp or flax.
b. **Particulates**: Micro or nanoparticles of materials such as silica, alumina, or CNTs.
c. **Whiskers**: Single-crystal ceramic whiskers.
d. **Fillers**: Micron-sized materials like glass beads, clay, or talc.

8.5.3   Advantages of Epoxy Composites

Advantages of epoxy composites include:

- **High Strength-to-Weight Ratio**: Epoxy composites offer excellent strength and stiffness while being relatively lightweight, making them suitable for weight-sensitive applications like aerospace and automotive industries.
- **Tailorable Properties**: By choosing different reinforcement materials and adjusting their proportions, engineers can customize epoxy composites to meet specific performance requirements for various applications.
- **Corrosion Resistance**: Epoxy composites are inherently resistant to corrosion, making them ideal for applications in harsh environments, such as marine structures.
- **Fatigue Resistance**: Epoxy composites show excellent fatigue resistance, making them suitable for load-bearing components subject to cyclic loading.
- **Design Flexibility**: Epoxy composites can be moulded into complex shapes, offering design flexibility for intricate components [56].

8.5.4   Applications of Epoxy Composites

Applications of epoxy composites include:

- **Aerospace**: Used in aircraft components, such as wings, fuselage, and interior parts, to reduce weight and enhance structural integrity.
- **Automotive**: Employed in body panels, suspension components, and interior parts to enhance fuel efficiency and performance.
- **Marine**: Used for boat hulls, decks, and marine structures to improve durability and resistance to water-related damage.
- **Sports and Recreation**: Utilized in bicycles, golf clubs, tennis rackets, and skis to improve performance and durability.
- **Electronics**: Employed in printed circuit boards (PCBs) and electronic encapsulation to provide electrical insulation and protection.

The epoxy composites offer a wide range of advantages and applications, making them indispensable materials in modern engineering and manufacturing industries.

Ongoing research and advancements in composite technology continue to expand their use and potential in diverse fields [57].

8.5.5 Properties of Epoxy Composites

The mechanical properties of epoxy composites are crucial factors that determine their performance and suitability for various engineering applications. These properties are influenced by the type and volume fraction of reinforcement materials used, the fabrication method, and the processing conditions [58]. The key mechanical properties of epoxy composites are discussed below:

- **Tensile Strength**: Tensile strength is the maximum stress a material can withstand before it fractures under tensile loading. Epoxy composites with high-strength reinforcements, such as carbon fibres or aramid fibres, exhibit enhanced tensile strength compared to unreinforced epoxy.
- **Compressive Strength**: Compressive strength refers to the ability of a material to withstand compressive loads before failure. Epoxy composites can have significant compressive strength, making them suitable for load-bearing applications.
- **Flexural Strength**: Flexural strength, also known as bending strength, is the ability of a material to resist deformation and failure under bending loads. Epoxy composites with strong and stiff reinforcements demonstrate improved flexural strength.
- **Shear Strength**: Shear strength is the resistance of a material to sliding or distortion along parallel planes. Epoxy composites often show good shear strength due to the strong interfacial bonding between the matrix and reinforcement.
- **Young's Modulus** (Elastic Modulus): Young's modulus is a measure of a material's stiffness and its ability to deform elastically under an applied load. Epoxy composites with high-stiffness reinforcements, like carbon fibres, have higher Young's modulus values than the unreinforced epoxy.
- **Poisson's Ratio**: Poisson's ratio characterizes the lateral strain that occurs in a material when it is subjected to uniaxial stress. For most epoxy composites, Poisson's ratio is relatively low.
- **Hardness**: Hardness is the resistance of a material to localized deformation, such as scratching or indentation. Epoxy composites often exhibit higher hardness than the matrix material alone due to the presence of the reinforcement.
- **Impact Strength**: Impact strength is the ability of a material to resist fracture under sudden impact or shock loading. The incorporation of certain reinforcements can improve the impact resistance of epoxy composites.
- **Fatigue Strength**: Fatigue strength refers to the resistance of a material to failure under cyclic loading. Epoxy composites can demonstrate excellent fatigue resistance, especially when reinforced with fibres [59].

It is important to note that the mechanical properties of epoxy composites can vary significantly depending on the type, alignment, and orientation of the reinforcement materials, as well as the manufacturing process used. Engineers carefully tailor these properties to meet specific application requirements, taking advantage of the versatility and tunability of epoxy composites for a wide range of engineering applications, including aerospace, automotive, marine, electronics, and sporting goods [60].

8.5.6 MANUFACTURING METHODS OF EPOXY COMPOSITES

The manufacturing methods of epoxy composites involve combining epoxy resin with reinforcement materials to create a homogeneous and strong material with enhanced mechanical and functional properties. The choice of manufacturing method depends on the desired properties of the composite, the type of reinforcement materials, and the intended application [61] Some common manufacturing methods for epoxy composites are as follows:

8.5.6.1 Hand Lay-up

In hand lay-up, layers of reinforcement materials (e.g., fiberglass, carbon fibres) are manually placed in a mould or on a tool surface. Epoxy resin is then applied to impregnate the reinforcement layers, ensuring good wetting and bonding between the resin and fibres. The composite is cured under controlled conditions, often at room temperature or elevated temperature, to achieve the desired properties.

8.5.6.2 Vacuum Bagging

Vacuum bagging is an extension of the hand lay-up method. After placing the reinforcement layers in the mould, a vacuum bag is sealed around the part to remove air and ensure uniform pressure during curing. The vacuum consolidates the layers, improves fibre-to-resin contact, and reduces voids in the composite.

8.5.6.3 Resin Transfer Moulding

RTM is a closed-mould process used for manufacturing complex composite parts with a high fibre volume fraction. Pre-cut reinforcement materials are placed in a mould, and the mould is closed before injecting epoxy resin into the cavity under pressure. The resin flows through the reinforcement, impregnating it completely, and then the composite is cured under controlled temperature and pressure.

8.5.6.4 Filament Winding

Filament winding is suitable for producing cylindrical or tubular composite structures, such as pipes or pressure vessels. Continuous fibres (e.g., carbon or glass) are wound onto a rotating mandrel, and epoxy resin is applied to impregnate the fibres as they are wound. The composite is then cured in an oven or through a controlled heating process.

8.5.6.5 Pultrusion

Pultrusion is a continuous manufacturing process where continuous fibres are pulled through a resin bath and then through a heated die to cure the composite. This method is suitable for producing composite profiles with a consistent cross-section, such as rods, tubes, and beams.

8.5.6.6   Compression Moulding

In compression moulding, pre-cut reinforcement materials are placed in a mould, and epoxy resin is added before closing the mould. The composite is cured under heat and pressure, ensuring the uniform distribution of the resin and reinforcement.

8.5.6.7   Injection Moulding

Injection moulding is commonly used for producing thermoplastic-based composites with short fibres or nanoparticles. The reinforcement is mixed with molten resin and then injected into a mould to form the desired shape.

Each of these manufacturing methods offers advantages and limitations, making them suitable for different applications and component geometries. Manufacturers select the appropriate method based on factors such as the complexity of the part, production volume, and desired mechanical properties of the epoxy composite [62].

8.6   CNT-REINFORCED EPOXY COMPOSITES

CNT-reinforced epoxy composites are advanced materials that combine CNTs with epoxy resin to create a high-performance composite material. The addition of CNTs as reinforcements imparts unique properties to the epoxy composites, making them attractive for various engineering applications. These composites are extensively studied due to their potential to revolutionize multiple industries. Key characteristics and features of CNT-reinforced epoxy composites as shown in Figure 8.3 and discussed below:

- **Enhanced Mechanical Properties**: CNTs are known for their exceptional mechanical strength and stiffness. When incorporated into epoxy composites, they reinforce the material, leading to improved tensile strength, flexural strength, and modulus.
- **High Aspect Ratio**: CNTs have a high aspect ratio, with lengths that can reach micrometres to millimetres while maintaining nanoscale diameters. This aspect ratio contributes to the reinforcement efficiency of the composites.
- **Lightweight**: Epoxy composites are inherently lightweight, and the addition of CNTs does not compromise this advantage. As a result, CNT-reinforced epoxy composites offer high strength-to-weight ratios.
- **Electrical Conductivity**: CNTs possess excellent electrical conductivity along their axial direction. When dispersed in epoxy, they can provide electrical conductivity, making the composites suitable for electromagnetic shielding or electrical applications.
- **Thermal Conductivity**: CNTs have high thermal conductivity, enabling efficient heat dissipation. This property is advantageous in applications where thermal management is crucial.
- **Improved Fracture Toughness**: The presence of CNTs can enhance the fracture toughness and impact resistance of epoxy composites, making them more resistant to crack propagation and impact loads.
- **Tailorable Properties**: The properties of CNT-reinforced epoxy composites can be tailored by adjusting the CNT content, dispersion, and alignment to suit specific application requirements [63].

MULTIPLE INDUSTRIAL KEY CHARACTERISTICS AND FEATURES OF CNT-REINFORCED EPOXY COMPOSITES

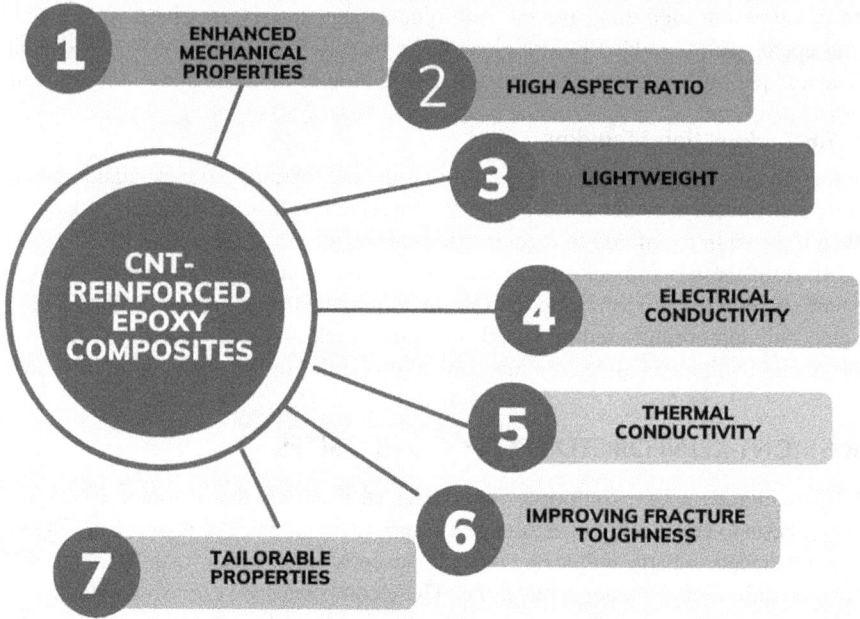

1 ENHANCED MECHANICAL PROPERTIES

2 HIGH ASPECT RATIO

3 LIGHTWEIGHT

CNT-REINFORCED EPOXY COMPOSITES

4 ELECTRICAL CONDUCTIVITY

5 THERMAL CONDUCTIVITY

6 IMPROVING FRACTURE TOUGHNESS

7 TAILORABLE PROPERTIES

FIGURE 8.3 Multiple industrial key characteristics and features of CNT reinforced epoxy composites.

8.6.1 APPLICATIONS OF CNT-REINFORCED EPOXY COMPOSITES

CNT-reinforced epoxy composites can be used in a wide variety of applications, including:

- **Aerospace and Aviation**: These composites are used in aircraft components to reduce weight while maintaining structural integrity.
- **Automotive**: In the automotive industry, CNT-reinforced epoxy composites find applications in lightweight vehicle components, leading to fuel efficiency and reduced emissions.
- **Electronics**: These composites are utilized in printed circuit boards (PCBs) and electrical connectors to provide mechanical strength and electrical conductivity.
- **Sports and Recreation**: CNT-reinforced epoxy composites are employed in sports equipment like tennis rackets, golf clubs, and bicycles, improving performance and durability.
- **Structural Reinforcements**: In civil engineering, CNT-reinforced epoxy composites can be used for strengthening and repair of concrete structures.
- **Energy Storage**: CNT-reinforced epoxy composites are investigated for applications in energy storage devices such as supercapacitors and batteries.

- **Biomedical Applications**: Due to their biocompatibility and mechanical properties, these composites have potential applications in biomedical devices and drug delivery systems.

However, challenges remain, such as ensuring uniform dispersion of CNTs, addressing agglomeration issues, and scaling up production processes. Ongoing research aims to overcome these challenges and unlock the full potential of CNT-reinforced epoxy composites for a wide range of industries [64].

8.6.2   MANUFACTURING METHODS OF CNT-REINFORCED EPOXY COMPOSITES

Reinforced epoxy composites involve combining CNTs with epoxy resin to create a homogeneous and high-performance composite material. The incorporation of CNTs as reinforcements enhances the mechanical, electrical, and thermal properties of the epoxy composites. Several methods are employed to produce these advanced materials, including:

8.6.2.1   Turning

Turning involves rotating the workpiece while a cutting tool removes material to achieve the desired shape and dimensions. Machining CNT-reinforced epoxy composites requires selecting appropriate cutting tools with high wear resistance due to the abrasive nature of CNTs.

8.6.2.2   Milling

Milling uses rotating multiple-point cutting tools to remove material from the workpiece. For machining CNT-reinforced epoxy composites, sharp and high-strength cutting tools are preferred to minimize tool wear.

8.6.2.3   Drilling

Drilling is used to create holes in the composite material. When drilling CNT-reinforced epoxy composites, proper cutting tool geometry and cooling techniques are essential to prevent delamination and ensure hole quality.

8.6.2.4   Abrasive Water Jet Machining

Abrasive Water Jet Machining (AWJM) utilizes a high-pressure jet of abrasive particles mixed with water to remove material from the workpiece. This process is particularly useful for cutting and shaping CNT-reinforced epoxy composites without generating significant heat and thermal damage.

8.6.2.5   Laser Machining

Laser machining involves using a high-energy laser beam to vaporize or melt the material, creating precise cuts or shapes. Laser machining can be advantageous for machining CNT-reinforced epoxy composites due to its accuracy and minimal thermal effects when used properly.

8.6.2.6   Electrical Discharge Machining (EDM)

EDM employs electrical discharges to remove material from the workpiece. This method is applicable for machining complex shapes in CNT-reinforced epoxy composites without causing mechanical stress.

8.6.2.7   Ultrasonic Machining

Ultrasonic machining uses ultrasonic vibrations in a slurry to remove material from the workpiece. This process can be effective for machining CNT-reinforced epoxy composites with minimal damage and improved surface finish.

8.6.3   CHALLENGES IN MACHINING CNT

Reinforced epoxy composites include tool wear, surface integrity, thermal effects, and the potential for CNT agglomeration. To overcome these challenges, selecting appropriate cutting parameters, cooling and lubrication techniques, and suitable cutting tools are crucial. Additionally, continuous research and advancements are necessary to improve the machinability of these advanced materials and enable their wider application in various engineering fields [65].

8.7   ADVANCED MACHINING TECHNIQUES OF CARBON NANOTUBES REINFORCED EPOXY COMPOSITES

Advanced machining techniques for CNTs reinforced epoxy composites aim to overcome the challenges associated with traditional machining methods and ensure efficient material removal while preserving the unique properties of the composites. These techniques utilize innovative approaches to enhance precision, reduce tool wear, minimize thermal effects, and improve surface integrity [66]. Some advanced machining techniques for CNT-reinforced epoxy composites include:

8.7.1   ULTRASONIC-ASSISTED MACHINING

Ultrasonic-assisted machining combines conventional machining with ultrasonic vibrations to improve cutting efficiency and reduce tool wear. Ultrasonic vibrations help break the bonding between CNTs and epoxy, making it easier to remove material.

8.7.2   CRYOGENIC MACHINING

Cryogenic machining involves using liquid nitrogen or other cryogenic gases as coolants during the machining process. The extreme cold temperature reduces the risk of thermal damage to the epoxy matrix and enhances tool life.

8.7.3   LASER-ASSISTED MACHINING

Laser-Assisted Machining (LAM) combines traditional machining with laser heating of the cutting zone to soften the epoxy matrix, making it easier to machine. The laser energy also reduces tool wear and improves the surface finish.

8.7.4   ELECTROCHEMICAL MACHINING

Electrochemical Machining (ECM) is a non-traditional machining method that uses electrochemical reactions to remove material from the workpiece. ECM can be advantageous for machining CNT-reinforced epoxy composites with complex shapes and minimal mechanical stress.

8.7.5   MICRO-EDM (ELECTRICAL DISCHARGE MACHINING)

Micro-EDM is a variant of EDM used for micro-machining applications. It can be utilized to create precise features in CNT-reinforced epoxy composites without causing significant thermal damage.

8.7.6   HIGH-SPEED MACHINING

High-speed machining involves using very high cutting speeds to improve material removal rates and reduce cutting forces. The technique can be adapted for machining CNT-reinforced epoxy composites with proper tooling and process control.

8.7.7   ABRASIVE FLOW MACHINING

Abrasive Flow Machining (AFM) employs a polymer media with abrasive particles to remove material from the workpiece. It can be used to achieve precise surface finishing of CNT-reinforced epoxy composites.

8.7.8   HYBRID MACHINING

Hybrid machining combines two or more machining processes, such as laser ablation followed by milling or ultrasonic-assisted turning. This approach can offer synergistic benefits and improve the overall machining efficiency and accuracy.

Each of these advanced machining techniques has specific advantages and limitations, and the selection depends on the specific requirements of the CNT-reinforced epoxy composite components. These techniques represent ongoing research and development efforts to optimize the machining process for advanced materials and enable their broader applications in industries ranging from aerospace and automotive to electronics and beyond [67].

8.7.9   CHALLENGES IN MACHINING CNT-REINFORCED EPOXY COMPOSITES

Machining CNTs reinforced epoxy composites present several challenges due to the unique properties and structure of these advanced materials. These challenges can impact the machining process and the quality of the machined components. Some of the key challenges are as follows and shown in Figure 8.4.

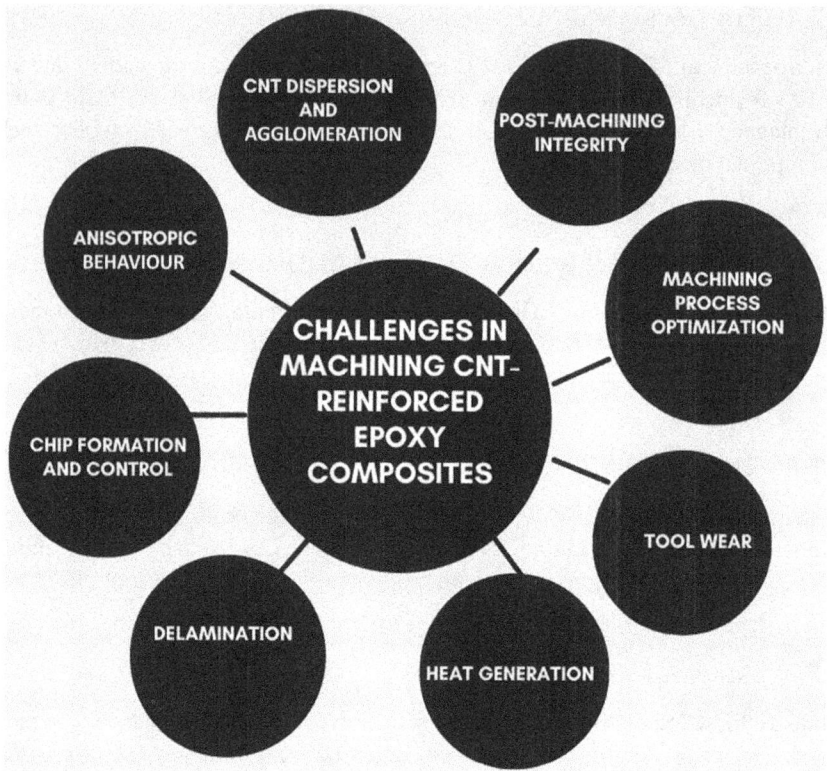

FIGURE 8.4   Challenges in machining CNT-reinforced epoxy composites.

- **Anisotropic Behavior**: CNT-reinforced epoxy composites often exhibit anisotropic behavior, where the mechanical properties vary with the orientation of the CNTs. This anisotropy can lead to non-uniform material removal during machining, resulting in dimensional inaccuracies and surface irregularities.
- **CNT Dispersion and Agglomeration**: Achieving a uniform dispersion of CNTs within the epoxy matrix is challenging. CNTs tend to agglomerate, forming clusters that affect material properties and complicate machining. These agglomerations can lead to uneven wear on cutting tools and poor surface finish.
- **Tool Wear**: The presence of CNTs in the composite can cause higher tool wear due to the high hardness and abrasive nature of these nanotubes. Tool wear reduces machining efficiency and can lead to increased costs and tool replacement.
- **Heat Generation**: Machining CNT-reinforced epoxy composites can generate significant heat due to the high thermal conductivity of CNTs. Excessive heat can cause thermal degradation of the epoxy matrix, leading to changes in material properties and affecting the dimensional accuracy of the machined components.

- **Delamination**: Delamination at the composite layers' interfaces can occur during machining, leading to reduced structural integrity and lower mechanical properties of the final component.
- **Chip Formation and Control**: The presence of CNTs can influence chip formation during machining, affecting chip size, shape, and evacuation. Improper chip control can cause workpiece damage and adversely impact surface finish.
- **Post-Machining Integrity**: After machining, the microstructure and mechanical properties of the CNT-reinforced epoxy composites may be altered. It is crucial to assess the post-machining integrity to ensure that the desired properties are maintained.
- **Machining Process Optimization**: Determining the optimal cutting parameters, tool materials, and machining strategies for CNT-reinforced epoxy composites is complex due to their unique properties. Extensive experimentation and analysis are necessary to develop efficient and effective machining processes.

Addressing these challenges requires a thorough understanding of the machinability of CNT-reinforced epoxy composites and the development of specialized machining techniques and tools. Ongoing research and advancements in materials science and machining technologies are essential to overcome these challenges and fully exploit the potential of CNT-reinforced epoxy composites in various engineering applications [68].

8.8   PRACTICAL IMPLICATIONS

The machining of carbon nanotubes CNTs has several practical implications for industries and research fields. These implications arise from the unique properties and challenges associated with these advanced materials during machining processes. Here are some practical implications:

8.8.1   IMPROVED MATERIAL SELECTION AND DESIGN

Machining considerations play a significant role in material selection and component design. Engineers need to carefully evaluate the machinability of CNT-reinforced epoxy composites while considering their mechanical, electrical, and thermal properties. This evaluation can influence the design of components to ensure optimal performance and manufacturability.

8.8.2   SPECIALIZED TOOLING AND CUTTING PARAMETERS

To overcome challenges like tool wear and CNT agglomeration, specialized cutting tools and cutting parameters must be developed and employed. Practical implications include investing in advanced tool materials, coatings, and tool geometries that are better suited for machining CNT-reinforced epoxy composites.

8.8.3   Machining Process Optimization

The complexity of machining CNT-reinforced epoxy composites demands systematic process optimization. Researchers and manufacturers must conduct extensive experimentation and analysis to identify the most efficient machining strategies, which can lead to cost reduction and improved productivity [69].

8.8.4   Advanced Machining Techniques

Practical implications involve the adoption of advanced machining techniques such as ultrasonic-assisted machining, cryogenic machining, and laser-assisted machining. Integrating these techniques into manufacturing can enhance productivity, minimize thermal damage, and improve surface finish.

8.8.5   Enhanced Quality Control

The presence of CNTs in the composite may alter its microstructure and properties during machining. Practical implications include the need for stringent quality control measures to ensure the machined components meet the required specifications and maintain the desired mechanical performance.

8.8.6   Application-Specific Solutions

Different industries may have specific requirements for CNT-reinforced epoxy composites. Practical implications involve tailoring machining processes to meet application-specific demands, such as aerospace, automotive, electronics, or medical devices.

8.8.7   Research and Collaboration

The machining of CNT-reinforced epoxy composites demands continuous research and collaboration between material scientists, engineers, and manufacturers. Practical implications include fostering interdisciplinary collaboration to address challenges and develop innovative solutions.

8.8.8   Economic and Environmental Impact

The successful machining of CNT-reinforced epoxy composites can lead to economic benefits by enabling the production of high-performance components. Additionally, the adoption of advanced machining techniques can reduce waste, energy consumption, and environmental impact.

The machining of CNT-reinforced epoxy composites has practical implications across various aspects of engineering and manufacturing. These implications call for continuous research, innovation, and collaboration to fully realize the potential of these advanced materials and integrate them into practical applications across industries [70].

8.9 SOME EXAMPLES

8.9.1 MACHINING STRATEGIES FOR AEROSPACE COMPONENTS

Investigate the machining strategies, tool materials, and cutting parameters for machining CNT-reinforced epoxy composite components used in aerospace applications. Focus on achieving dimensional accuracy, surface finish, and minimizing tool wear to meet stringent aerospace quality requirements [71].

8.9.2 OPTIMIZATION OF HYBRID MACHINING TECHNIQUES

Examine the benefits of combining laser-assisted machining with traditional milling or turning to machine CNT-reinforced epoxy composites. Evaluate the impact of laser power, cutting speed, and feed rate on cutting forces, tool life, and surface integrity.

8.9.3 CRYOGENIC MACHINING FOR HIGH-VOLUME PRODUCTION

Assess the viability of cryogenic machining as a cost-effective solution for high-volume production of CNT-reinforced epoxy composite parts. Analyze the influence of cryogenic cooling on tool wear, surface quality, and overall process economics.

8.9.4 REAL-TIME MONITORING AND ADAPTIVE MACHINING

Implement real-time monitoring and adaptive machining strategies for CNT-reinforced epoxy composites. Evaluate the effectiveness of sensor-based feedback in adjusting cutting parameters to optimize machining efficiency and reduce variability.

8.9.5 SUSTAINABLE MACHINING PRACTICES

Investigate sustainable machining practices for CNT-reinforced epoxy composites, such as the use of eco-friendly coolants and lubricants. Assess their impact on energy consumption, waste reduction, and the overall environmental footprint.

8.9.6 MACHINING OF BIOMEDICAL IMPLANTS

Examine the challenges and opportunities of machining CNT-reinforced epoxy composites for biomedical implant applications. Evaluate the effects of machining on biocompatibility, surface roughness, and mechanical properties critical for implant success.

8.9.7 MULTI-AXIS MACHINING OF COMPLEX GEOMETRIES

Explore multi-axis machining techniques for producing complex geometries in CNT-reinforced epoxy composites. Evaluate the feasibility of machining intricate parts, such as turbine blades or custom medical devices, with high precision and surface quality.

These hypothetical case study topics aim to showcase the practical implications of machining CNT-reinforced epoxy composites across different industries and research areas. As more research is conducted and industries gain experience with these materials, more real-world case studies are expected to emerge, shedding light on the best practices and challenges in machining these advanced materials [72].

8.10   RECOMMENDATIONS FOR FUTURE WORK

Future work in the machining of CNTs reinforced epoxy composites should focus on addressing the existing challenges and advancing the capabilities of machining processes for these advanced materials. Here are some recommendations for future research and development in this area:

- **CNT Dispersion Techniques**: Investigate and develop improved CNT dispersion techniques within the epoxy matrix to achieve better uniformity and dispersion. Effective dispersion methods will lead to reduced agglomeration and improved machinability.
- **Tool Material Development**: Explore and develop advanced tool materials, coatings, and geometries specifically designed for machining CNT-reinforced epoxy composites. High-performance tooling can enhance tool life, reduce wear, and improve machining efficiency.
- **Machining Process Optimization**: Conduct extensive research to optimize machining parameters, cutting speeds, feeds, and depths of cut for different machining techniques. Utilize experimental and computational approaches to determine the optimal conditions that minimize tool wear and surface damage while ensuring dimensional accuracy.
- **Adaptive Machining Strategies**: Implement adaptive machining strategies that adjust cutting parameters in real-time based on feedback from sensors or monitoring systems. Adaptive machining can help adapt to variations in material properties, CNT distribution, and machining conditions, leading to improved process control.
- **Hybrid Machining**: Investigate the integration of multiple machining processes, such as combining laser machining with conventional machining or ultrasonic-assisted machining. Hybrid machining can leverage the advantages of different techniques and offer improved machining efficiency and precision.
- **Modeling and Simulation**: Develop advanced modeling and simulation tools to predict and understand the behavior of CNT-reinforced epoxy composites during machining. Computational simulations can aid in process planning, predicting tool wear, and optimizing machining parameters.
- **Surface Integrity and Quality Control**: Research on the effects of machining on the surface integrity and mechanical properties of CNT-reinforced epoxy composites. Develop non-destructive testing methods and inspection techniques to ensure the quality and integrity of machined components.
- **Sustainability and Environmental Impact**: Investigate sustainable machining practices for CNT-reinforced epoxy composites that reduce energy

consumption and waste generation. This includes exploring eco-friendly coolants, lubricants, and recycling of machining by-products.

- **Real-World Applications**: Conduct case studies and practical applications of machined CNT-reinforced epoxy composite components in industries such as aerospace, automotive, electronics, and renewable energy. Validate the performance and reliability of machined parts in real-world settings.
- **Collaboration and Knowledge Sharing**: Foster collaboration between researchers, industries, and academic institutions to share knowledge, best practices, and findings related to the machining of CNT-reinforced epoxy composites. Collaborative efforts can accelerate progress and lead to innovative solutions.

By addressing these recommendations, future work in the machining of CNT-reinforced epoxy composites can contribute to unlocking the full potential of these advanced materials, enabling their widespread adoption in high-performance engineering applications.

8.11   CONCLUDING REMARKS

The machining CNTs reinforced epoxy composites present both opportunities and challenges for the manufacturing industry. These advanced materials offer exceptional mechanical, electrical, and thermal properties, making them highly desirable for a wide range of engineering applications. Machining processes play a crucial role in shaping and finishing these composites to meet specific design requirements.

However, the machining of CNT-reinforced epoxy composites is not without its difficulties. The anisotropic behavior, CNT dispersion, agglomeration, tool wear, heat generation, delamination, and post-machining integrity are significant challenges that need to be addressed. Researchers and manufacturers must work together to develop specialized machining techniques and optimize cutting parameters to achieve precise and efficient material removal while preserving the composite's unique properties.

Advanced machining techniques, such as ultrasonic-assisted machining, cryogenic machining, laser-assisted machining, and others, show promise in mitigating some of these challenges. These methods aim to reduce tool wear, control thermal effects, and improve surface finish during machining processes.

To fully harness the potential of CNT-reinforced epoxy composites, ongoing research and development efforts are necessary. Improved CNT dispersion techniques, innovative tool materials, and a deeper understanding of the composite's behavior during machining are vital to enable the practical application of these advanced materials in industries like aerospace, automotive, electronics, and more.

The machining of CNT-reinforced epoxy composites is a dynamic and evolving field. By overcoming the challenges and optimizing the machining processes, engineers and manufacturers can unlock the full potential of these composites, contributing to advancements in materials science and paving the way for innovative and high-performance engineering solutions in various sectors.

REFERENCES

1. Liu, B., et al., Stochastic integrated machine learning based multiscale approach for the prediction of the thermal conductivity in carbon nanotube reinforced polymeric composites. *Composites Science and Technology* 224 (2022): p. 109425.
2. Uppal, A.S., Sharma, A., Babbar, A., Singh, K. and Singh, A.K., Minimum quality lubricant (MQL) for ultraprecision machining of titanium nitride-coated carbide inserts: Sustainable Manufacturing process. *International Journal on Interactive Design and Manufacturing (IJIDeM)* (2023): pp. 1–12.
3. Bansal, S., Kaushal, S., Mago, J., Gupta, D., Jain, V., Babbar, A. et al., Effect of variation of WC reinforcement on metallurgical and cavitation erosion behavior of microwave processed NiCrSiC-WC composites clads. *Proceedings of the Institution of Mechanical Engineers, Part C: Journal of Mechanical Engineering Science* (2023): p. 09544062 2311645.
4. Babbar, V. Jain, D. Gupta, K. Goyal, C. Prakash, K. Saxena et al., Investigation of infrared thermography of cortical bone grinding in neurosurgery. *Advances in Science and Technology Research Journal* 17 (2023): pp. 116–123.
5. Tian, Y., Tian, C., Han, J., Babbar, A. and Liu, B., Characteristics of grinding force and Kevlar deformation of novel body-armor-like abrasive tool. *The International Journal of Advanced Manufacturing Technology* 122 (2022): pp. 2019–2030.
6. Gu, Z., Tian, Y., Han, J., Wei, C., Babbar, A. and Liu, B., Characteristics of high-shear and low-pressure grinding for Inconel718 alloy with a novel super elastic composite abrasive tool. *The International Journal of Advanced Manufacturing Technology* 123 (2022): pp. 345–355.
7. Sharma, A. Babbar, Y. Tian, B.P. Pathri, M. Gupta and Singh, R., Machining of ceramic materials: A state-of-the-art review. *International Journal on Interactive Design and Manufacturing (IJIDeM)* 17 (2022): pp. 2891–2911.
8. Babbar, V. Jain, D. Gupta, D. Agrawal, C. Prakash, S. Singh et al., Experimental analysis of wear and multi-shape burr loading during neurosurgical bone grinding. *Journal of Materials Research and Technology* 12 (2021): pp. 15–28.
9. Babbar, V. Jain, D. Gupta and Agrawal, D., Finite element simulation and integration of CEM43°C and Arrhenius Models for ultrasonic-assisted skull bone grinding: A thermal dose model. *Medical Engineering & Physics* 90 (2021): pp. 9–22.
10. Babbar, A., Jain, V. and Gupta, D., In vivo evaluation of machining forces, torque, and bone quality during skull bone grinding. *Proceedings of the Institution of Mechanical Engineers, Part H: Journal of Engineering in Medicine* 234 (2020): pp. 626–638.
11. Babbar, V. Jain and Gupta, D., Thermogenesis mitigation using ultrasonic actuation during bone grinding: A hybrid approach using CEM43°C and Arrhenius model. *Journal of the Brazilian Society of Mechanical Sciences and Engineering* 41 (2019): p. 401.
12. Babbar, A. Sharma and Singhm, P., Multi-objective optimization of magnetic abrasive finishing using grey relational analysis. *Materials Today: Proceedings* 50 (2022): pp. 570–575.
13. Sharma, M. Kalsia, A.S. Uppal, A. Babbar and Dhawan, V., Machining of hard and brittle materials: A comprehensive review. *Materials Today: Proceedings* 50 (2022): pp. 1048–1052.
14. Prakash, V. Kumar, A. Mistri, A.S. Uppal, A. Babbar, B.P. Pathri et al., Investigation of functionally graded adherents on failure of socket joint of FRP composite tubes. *Materials* 14 (2021): pp. 6365.
15. Sharma, V. Kumar, A. Babbar, V. Dhawan, K. Kotecha and Prakash, C., Experimental investigation and optimization of electric discharge machining process parameters using Grey-fuzzy-based hybrid techniques. *Materials* 14 (2021): pp. 5820.

16. Babbar, V. Jain, Gupta, D. and Prakash, C., Experimental investigation and parametric optimization of neurosurgical bone grinding under bio-mimic environment. *Surface Review and Letters* 30 (2023): p. 2141005.

17. Singh, G., Babbar, A., Jain, V. and Gupta, D., Comparative statement for diametric delamination in drilling of cortical bone with conventional and ultrasonic assisted drilling techniques. *Journal of Orthopaedics* 25 (2021): pp. 53–58.

18. Singh, S., Prakash, C., Pramanik, A., Basak, A., Shabadi, R., Królczyk, G. et al., Magneto-rheological fluid assisted abrasive nanofinishing of β-phase Ti-Nb-Ta-Zr alloy: Parametric appraisal and corrosion analysis. *Materials* 13 (2020): p. 5156.

19. Sharma, V. Jain, D. Gupta and Babbar, A., A Review Study on Miniaturization, in Chander Prakash, Sunpreet Singh and J. Paulo Davim (Eds.), *Advanced Manufacturing and Processing Technology* (First edition, pp. 111–131), CRC Press, Boca Raton, FL, [2021] 2020.

20. Babbar, V. Jain, D. Gupta, C. Prakash and Sharma, A., Fabrication and Machining Methods of Composites for Aerospace Applications, in Chander Prakash, Sunpreet Singh and J. Paulo Davim (Eds.), *Characterization, Testing, Measurement, and Metrology* (First edition, pp. 109–124). CRC Press, Boca Raton, 2020.

21. Babbar, V. Jain, D. Gupta, C. Prakash, S. Singh and Sharma, A., Effect of Process Parameters on Cutting Forces and Osteonecrosis for Orthopedic Bone Drilling Applications, in Chander Prakash, Sunpreet Singh and J. Paulo Davim (Eds.), *Characterization, Testing, Measurement, and Metrology* (First edition, pp. 93–108). CRC Press, Boca Raton, 2020.

22. Babbar, V. Jain, D. Gupta and Sharma, A., Fabrication of Microchannels using Conventional and Hybrid Machining Processes, in Rupinder Singh and J. Paulo Davim (Eds.), *Non-Conventional Hybrid Machining Processes* (First edition, pp. 37–51). CRC Press, Boca Raton, 2020.

23. Sharma, V. Grover, A. Babbar and Rani, R., A Trending Nonconventional Hybrid Finishing/Machining Process, in Rupinder Singh and J. Paulo Davim (Eds.), *Non-Conventional Hybrid Machining Processes* (First edition, pp. 79–93). CRC Press, Boca Raton, 2020.

24. Babbar, C. Prakash, S. Singh, M.K. Gupta, M. Mia and Pruncu, C.I., Application of hybrid nature-inspired algorithm: Single and bi-objective constrained optimization of magnetic abrasive finishing process parameters. *Journal of Materials Research and Technology* 9 (2020): pp. 7961–7974.

25. Baraiya, R., Babbar, A., Jain, V. and Gupta, D., In-situ simultaneous surface finishing using abrasive flow machining via novel fixture. *Journal of Manufacturing Processes* 50 (2020): pp. 266–278.

26. Babbar, A., Sharma, A., Jain, V. and Jain, A.K., Rotary ultrasonic milling of C/SiC composites fabricated using chemical vapor infiltration and needling technique. *Materials Research Express* 6 (2019): p. 085607.

27. Babbar, A., Jain, V. and Gupta, D., Neurosurgical Bone Grinding, in Chander Prakash, Sunpreet Singh, Rupinder Singh, Seeram Ramakrishna, B.S. Pabla, Sanjeev Puri and M.S. Uddin (Eds.), *Biomanufacturing* (pp. 137–155). Springer International Publishing, Cham, 2019.

28. Sharma, A., Babbar, A., Jain, V. and Gupta, D., Enhancement of surface roughness for brittle material during rotary ultrasonic machining. *MATEC Web of Conferences* 249 (2018): p. 01006.

29. Panchagnula, K.K., et al., CoCoSo method-based optimization of cryogenic drilling on multi-walled carbon nanotubes reinforced composites. *International Journal on Interactive Design and Manufacturing (IJIDeM)* 17(1) (2023): pp. 279–297.

30. Liu, B., et al., Stochastic full-range multiscale modeling of thermal conductivity of polymeric carbon nanotubes composites: A machine learning approach. *Composite Structures* 289 (2022): p. 115393.

31. Xu, Z., et al., Improving the mechanical properties of carbon nanotubes reinforced aluminum matrix composites by heterogeneous structural design. *Composites Communications* 29 (2022): p. 101050.

32. Jongvivatsakul, P., et al., Enhancing bonding behavior between carbon fiber-reinforced polymer plates and concrete using carbon nanotube reinforced epoxy composites. *Case Studies in Construction Materials* 17 (2022): p. e01407.

33. Darıcık, F., et al., Carbon nanotube (CNT) modified carbon fiber/epoxy composite plates for the PEM fuel cell bipolar plate application. *International Journal of Hydrogen Energy* 48(3) (2023): pp. 1090–1106.

34. Raj, S.O.N. and Prabhu, S., AFM analysis on surface roughness of single crystal silicon machined with carbon nanotubes reinforced composite micro grinding wheel. *Silicon* 14(12) (2022): pp. 7305–7320.

35. Doğan, K., et al., Dispersion mechanism-induced variations in microstructural and mechanical behavior of CNT-reinforced aluminum nanocomposites. *Archives of Civil and Mechanical Engineering* 22(1) (2022): p. 55.

36. Li, X., et al., Wear behavior of the uniformly dispersed carbon nanotube reinforced 6061Al composite fabricated by milling combined with powder metallurgy. *Acta Metallurgica Sinica (English Letters)* 35(11) (2022): pp. 1765–1776.

37. Wang, X., et al., Ultimate tensile behavior of short single-wall carbon nanotube/epoxy composites and the reinforced mechanism. *Polymer Composites* 44(4) (2023): pp. 2545–2556.

38. Mishra, L., Mahapatra, T.R., and Mishra, D., Performance evaluation and sustainability assessment in laser micro-drilling of carbon nanotube-reinforced polymer matrix composite using MOORA and whale optimization algorithm. *Process Integration and Optimization for Sustainability* 6(3) (2022): pp. 603–620.

39. Nateq, B., et al., Interfacial structure-property relationship in a carbon nanotube-reinforced aluminum alloy matrix composite fabricated by an advanced method. *Materials Characterization* 203 (2023): pp. 113159.

40. Mei, L., et al., Multiscale carbon nanotube-carbon fiber reinforcement for advanced epoxy composites with high interfacial strength. *Polymers and Polymer Composites* 19(2–3) (2011): pp. 107–112.

41. Cho, J. and Daniel, I., Reinforcement of carbon/epoxy composites with multi-wall carbon nanotubes and dispersion enhancing block copolymers. *Scripta Materialia* 58(7) (2008): pp. 533–536.

42. Ci, L. and Bai, J., The reinforcement role of carbon nanotubes in epoxy composites with different matrix stiffness. *Composites Science and Technology* 66(3–4) (2006): pp. 599–603.

43. Singh, N.P., Gupta, V., and Singh, A.P., Graphene and carbon nanotube reinforced epoxy nanocomposites: A review. *Polymer* 180 (2019): pp. 121724.

44. Rahmanian, S., et al., Mechanical characterization of epoxy composite with multiscale reinforcements: Carbon nanotubes and short carbon fibers. *Materials & Design* 60 (2014): pp. 34–40.

45. Islam, M.E., et al., Characterization of carbon fiber reinforced epoxy composites modified with nanoclay and carbon nanotubes. *Procedia Engineering* 105 (2015): pp. 821–828.

46. Kumar, R., et al., Bibliometric analysis of specific energy consumption (SEC) in machining operations: A sustainable response. *Sustainability* 13(10) (2021): pp. 5617.

47. Rani, S., et al., Security and privacy challenges in the deployment of cyber-physical systems in smart city applications: State-of-art work. *Materials Today: Proceedings* 62 (2022): pp. 4671–4676.

48. Sidhu, A.S., et al., Prioritizing energy-intensive machining operations and gauging the influence of electric parameters: An industrial case study. *Energies* 14(16) (2021): p. 4761.
49. Kumar, R., et al., Analysis the effects of process parameters in EN24 alloy steel during CNC turning by using MADM. *International Journal of Innovative Research in Science, Engineering and Technology* 2 (2013): pp. 1131–1145.
50. Hassan, A., et al., Machinability investigation in electric discharge machining of carbon fiber reinforced composites for aerospace applications. *Polymer Composites* 43(11) (2022): pp. 7773–7788.
51. Zhou, L., et al., Effect of vertical aligned carbon nanofiber/carbon nanotube on the mechanical, electrical and electromagnetic interference shielding properties of epoxy foams. *Composites Communications* 35 (2022): p. 101325.
52. Geng, H., et al., Matrix effect on strengthening behavior of carbon nanotubes in aluminum matrix composites. *Materials Characterization* 195 (2023): p. 112484.
53. Kumar, R., et al., Hand and abrasive flow polished tungsten carbide die: Optimization of surface roughness, polishing time and comparative analysis in wire drawing. *Materials* 15(4) (2022): p. 1287.
54. Channi, A.S., et al., Tool wear rate during electrical discharge machining for aluminium metal matrix composite prepared by squeeze casting: A prospect as a biomaterial. *Journal of Electrochemical Science and Engineering* 13(1) (2023): pp. 149–162.
55. Kumar, R. and Singh, S., Abrasive flow polishing of tungsten carbide wire drawing die. *International Journal of Applied Engineering Research* 6(4) (2011): pp. 499–510.
56. Cheng, Q., et al., Fabrication and properties of aligned multiwalled carbon nanotube-reinforced epoxy composites. *Journal of Materials Research* 23(11) (2008): pp. 2975–2983.
57. Chu, K., et al., Fabrication and effective thermal conductivity of multi-walled carbon nanotubes reinforced Cu matrix composites for heat sink applications. *Composites Science and Technology* 70(2) (2010): pp. 298–304.
58. Zhou, Y., et al., Experimental study on the thermal and mechanical properties of multi-walled carbon nanotube-reinforced epoxy. *Materials Science and Engineering: A* 452 (2007): pp. 657–664.
59. Olifirov, L., Kaloshkin, S., and Zhang, D., Study of thermal conductivity and stress-strain compression behavior of epoxy composites highly filled with Al and Al/f-MWCNT obtained by high-energy ball milling. *Composites Part A: Applied Science and Manufacturing* 101 (2017): pp. 344–352.
60. Yan, L., et al., Friction and wear properties of aligned carbon nanotubes reinforced epoxy composites under water lubricated condition. *Wear* 308(1–2) (2013): pp. 105–112.
61. Kumar, D. and Singh, K., Investigation of delamination and surface quality of machined holes in drilling of multiwalled carbon nanotube doped epoxy/carbon fiber reinforced polymer nanocomposite. *Proceedings of the Institution of Mechanical Engineers, Part L: Journal of Materials: Design and Applications* 233(4) (2019): pp. 647–663.
62. Sharma, K. and Shukla, M., Three-phase carbon fiber amine functionalized carbon nanotubes epoxy composite: Processing, characterisation, and multiscale modeling. *Journal of Nanomaterials* 2014 (2014): pp. 2–2.
63. Wernik, J. and Meguid, S., On the mechanical characterization of carbon nanotube reinforced epoxy adhesives. *Materials & Design* 59 (2014): pp. 19–32.
64. Rahman, A., et al., A machine learning framework for predicting the shear strength of carbon nanotube-polymer interfaces based on molecular dynamics simulation data. *Composites Science and Technology* 207 (2021): p. 108627.

65. Thostenson, E.T. and Chou, T.-W., Processing-structure-multi-functional property relationship in carbon nanotube/epoxy composites. *Carbon* 44(14) (2006): pp. 3022–3029.

66. Kumar, D., Singh, K.K. and Zitoune, R., Impact of the carbon nanotube reinforcement in glass/epoxy polymeric nanocomposite on the quality of fiber laser drilling. *Proceedings of the Institution of Mechanical Engineers, Part B: Journal of Engineering Manufacture* 232(14) (2018): pp. 2533–2546.

67. Kim, K.T., et al., Microstructures and tensile behavior of carbon nanotube reinforced Cu matrix nanocomposites. *Materials Science and Engineering: A* 430(1–2) (2006): pp. 27–33.

68. Loos, M., et al., Enhanced fatigue life of carbon nanotube-reinforced epoxy composites. *Polymer Engineering & Science* 52(9) (2012): pp. 1882–1887.

69. Liu, X., et al., Microstructure evolution and mechanical properties of carbon nanotubes reinforced Al matrix composites. *Materials Characterization* 133 (2017): pp. 122–132.

70. Suresha, B., et al., Effect of carbon nanotubes reinforcement on mechanical properties of aramid/epoxy hybrid composites. *Materials Today: Proceedings* 43 (2021): pp. 1478–1484.

71. Wicks, S.S., de Villoria, R.G. and Wardle, B.L., Interlaminar and intralaminar reinforcement of composite laminates with aligned carbon nanotubes. *Composites Science and Technology* 70(1) (2010): pp. 20–28.

72. Rajak, D.K., et al., Recent progress of reinforcement materials: A comprehensive overview of composite materials. *Journal of Materials Research and Technology* 8(6) (2019): pp. 6354–6374.

Index